T0291683

CAMBRIDGE LIBRARY COLLECTION

Books of enduring scholarly value

Mathematical Sciences

From its pre-historic roots in simple counting to the algorithms powering modern desktop computers, from the genius of Archimedes to the genius of Einstein, advances in mathematical understanding and numerical techniques have been directly responsible for creating the modern world as we know it. This series will provide a library of the most influential publications and writers on mathematics in its broadest sense. As such, it will show not only the deep roots from which modern science and technology have grown, but also the astonishing breadth of application of mathematical techniques in the humanities and social sciences, and in everyday life.

The Mysterious Universe

Originating from the Rede Lecture delivered at the University of Cambridge in November 1930, this book is based upon the conviction that the teachings and findings of astronomy and physical science are destined to produce an immense change on our outlook on the universe as a whole, and on views about the significance of human life. The author contends that the questions at issue are ultimately one for philosophical discussion, but that before philosophers can speak, science should present ascertained facts and provisional hypotheses. The book is therefore written with these thoughts in mind while broadly presenting the fundamental physical ideas and findings relevant for a wider philosophical inquiry.

Cambridge University Press has long been a pioneer in the reissuing of out-of-print titles from its own backlist, producing digital reprints of books that are still sought after by scholars and students but could not be reprinted economically using traditional technology. The Cambridge Library Collection extends this activity to a wider range of books which are still of importance to researchers and professionals, either for the source material they contain, or as landmarks in the history of their academic discipline.

Drawing from the world-renowned collections in the Cambridge University Library, and guided by the advice of experts in each subject area, Cambridge University Press is using state-of-the-art scanning machines in its own Printing House to capture the content of each book selected for inclusion. The files are processed to give a consistently clear, crisp image, and the books finished to the high quality standard for which the Press is recognised around the world. The latest print-on-demand technology ensures that the books will remain available indefinitely, and that orders for single or multiple copies can quickly be supplied.

The Cambridge Library Collection will bring back to life books of enduring scholarly value (including out-of-copyright works originally issued by other publishers) across a wide range of disciplines in the humanities and social sciences and in science and technology.

The Mysterious Universe

James Jeans

CAMBRIDGE UNIVERSITY PRESS

Cambridge, New York, Melbourne, Madrid, Cape Town, Singapore,
São Paolo, Delhi, Dubai, Tokyo

Published in the United States of America by Cambridge University Press, New York

www.cambridge.org
Information on this title: www.cambridge.org/9781108005661

This edition first published 1930
This digitally printed version 2009

ISBN 978-1-108-00566-1

THE MYSTERIOUS UNIVERSE

PLATE I. THE DEPTHS OF SPACE

A Cluster of Nebulae in Coma Berenices. This is a photograph of a minute piece of the sky, taken with the largest telescope in existence (Mount Wilson, 100-inch). The majority of objects are nebulae, at a distance such that their light takes 50 million years to reach us. Each nebula contains some thousands of millions of stars, or the material for their formation. About two million such nebulae can be photographed in all, and there are probably millions of millions of others beyond the range of any telescope (see p. 57)

THE MYSTERIOUS UNIVERSE

by

SIR JAMES JEANS

O.M., F.R.S.

CAMBRIDGE

AT THE UNIVERSITY PRESS

1948

CAMBRIDGE UNIVERSITY PRESS
Cambridge, New York, Melbourne, Madrid, Cape Town, Singapore, São Paulo, Delhi

Cambridge University Press
The Edinburgh Building, Cambridge CB2 8RU, UK

Published in the United States of America by Cambridge University Press, New York

www.cambridge.org
Information on this title: www.cambridge.org/9780521054171

First published 1930
Reprinted (with further corrections) 1930
Second edition 1931
Reprinted (with corrections) 1933
Reprinted 1948
This digitally printed version 2008

A catalogue record for this publication is available from the British Library

ISBN 978-0-521-05417-1 hardback
ISBN 978-0-521-09001-8 paperback

CONTENTS

And now, I said, let me show in a figure how far our nature is enlightened or unenlightened:—Behold! human beings living in an underground cave, which has a mouth open towards the light and reaching all along the cave; here they have been from their childhood, and have their legs and necks chained so that they cannot move, and can only see before them, being prevented by the chains from turning round their heads. Above and behind them a fire is blazing at a distance, and between the fire and the prisoners there is a raised way; and you will see, if you look, a low wall built along the way, like the screen which marionette players have in front of them, over which they show the puppets.

I see.

And do you see, I said, men passing along the wall carrying all sorts of vessels, and statues and figures of animals made of wood and stone and various materials, which appear over the wall?...

You have shown me a strange image, and they are strange prisoners.

Like ourselves, I replied; and they see only their own shadows, or the other shadows which the fire throws on the opposite wall of the cave?

True, he said; how could they see anything but the shadows if they were never allowed to move their heads?

And of the objects which are being carried in like manner they would only see the shadows?

Yes, he said.

To them, I said, the truth would be literally nothing but the shadows of the images.

PLATO, *Republic*, Book VII

FOREWORD

The present book contains an expansion of the Rede Lecture delivered before the University of Cambridge in November 1930.

There is a widespread conviction that the new teachings of astronomy and physical science are destined to produce an immense change on our outlook on the universe as a whole, and on our views as to the significance of human life. The question at issue is ultimately one for philosophic discussion, but before the philosophers have a right to speak, science ought first to be asked to tell all she can as to ascertained facts and provisional hypotheses. Then, and then only, may discussion legitimately pass into the realms of philosophy.

With some such thoughts as these in my mind, I wrote the present book, obsessed by frequent doubts as to whether I could justify an addition to the great amount which has already been written on the subject. I can claim no special qualifications beyond the proverbially advantageous position of the mere onlooker; I am not a philosopher either by training or inclination, and for many years my scientific work has lain outside the arena of contending physical theories.

The first four chapters, which form the main part of the book, contain brief discussions, on very broad lines, of such scientific questions as seem to me to be of interest, and to provide useful material, for the discussion of the ultimate philosophical problem. As far as possible I have avoided overlapping my former book, *The Universe Around Us*, because I hope the present book may be read as a sequel to that. But an exception has been made in favour of material which is essential to the main argument, so as to make the present book complete in itself.

The last chapter stands on a different level. Every one may claim the right to draw his own conclusions from the facts presented by modern science. This chapter merely contains the interpretations which I, a stranger in the realms of philosophical thought, feel inclined to place on the scientific facts and hypotheses discussed in the main part of the book. Many will disagree with it—it was written to this end.

J. H. JEANS

DORKING, 1930

In preparing a second edition, I have tried to bring the scientific matter of the first four chapters up to date, and to remove all ambiguities from my argument. I found with regret that certain passages in the original book were liable to be misunderstood, misinterpreted, and even misquoted, in various unexpected ways. Some of these passages have been expunged, some rewritten and some amplified. Here and there new paragraphs, occasionally even whole pages, have been added in the hope of making the argument clearer.

J. H. JEANS

DORKING,
July 1st, 1931

Chapter I

THE DYING SUN

A few stars are known which are hardly bigger than the earth, but the majority are so large that hundreds of thousands of earths could be packed inside each and leave room to spare; here and there we come upon a giant star large enough to contain millions of millions of earths. And the total number of stars in the universe is probably something like the total number of grains of sand on all the seashores of the world. Such is the littleness of our home in space when measured up against the total substance of the universe.

This vast multitude of stars are wandering about in space. A few form groups which journey in company, but the majority are solitary travellers. And they travel through a universe so spacious that it is an event of almost unimaginable rarity for a star to come anywhere near to another star. For the most part each voyages in splendid isolation, like a ship on an empty ocean. In a scale model in which the stars are ships, the average ship will be well over a million miles from its nearest neighbour, whence it is easy to understand why a ship seldom finds another within hailing distance.

We believe, nevertheless, that some two thousand million years ago this rare event took place, and that a second star, wandering blindly through space, happened to come within hailing distance of the sun. Just as the sun and moon raise tides on the earth, so this second star must have raised tides on the surface of the sun. But they would be very different from the puny tides which the small mass of the moon raises in our oceans; a huge tidal wave must have travelled over the surface of the sun, ultimately forming a mountain of prodigious height, which would rise ever higher and higher as the cause of the disturbance came nearer and nearer. And, before the second star began to recede, its tidal pull had become so powerful that this mountain was torn to pieces and threw off small fragments of itself, much as the crest of a wave throws off spray. These small fragments have been circulating around their parent sun ever since. They are the planets, great and small, of which our earth is one.

The sun and the other stars we see in the sky are all intensely hot—far too hot for life to be able to obtain or retain a footing on them. So also no doubt were the ejected fragments of the sun when they were first thrown off. Gradually they cooled, until now they have but little intrinsic heat left, their warmth being derived almost entirely from the radiation which the sun pours down upon them. In course of time, we know not how, when, or why, one of these cooling fragments gave birth to life. It started in simple organisms whose vital capacities consisted of little beyond reproduction and death. But from these humble beginnings emerged a stream of life which, advancing through ever greater and greater complexity, has culminated in beings whose lives are largely centred in their emotions and ambitions, their aesthetic appreciations, and the religions in which their highest hopes and noblest aspirations lie enshrined.

Although we cannot speak with any certainty, it seems most likely that humanity came into existence in some such way as this. Standing on our microscopic fragment of a grain of sand, we attempt to discover the nature and purpose of the universe which surrounds our home in space and time. Our first impression is something akin to terror. We find the universe terrifying because of its vast meaningless distances, terrifying because of its inconceivably long vistas of time which dwarf human history to the twinkling of an eye, terrifying because of our extreme loneliness, and because of the material insignificance of our home in space —a millionth part of a grain of sand out of all the sea-sand in the world. But above all else, we find the universe terrifying because it appears to be indifferent to life like our own; emotion, ambition and achievement, art and religion all seem equally foreign to its plan. Perhaps indeed we ought to say it appears to be actively hostile to life like our own. For the most part, empty space is so cold that all life in it would be frozen; most of the matter in space is so hot as to make life on it impossible; space is traversed, and astronomical bodies continually bombarded, by radiation of a variety of kinds, much of which is probably inimical to, or even destructive of, life.

Into such a universe we have stumbled, if not exactly by mistake, at least as the result of what may properly be described as an accident. The use of such a word need not imply any surprise that our earth exists, for accidents will happen, and if the universe goes on for long enough, every conceivable accident is likely to happen in time. It was, I think, Huxley who said that six monkeys, set to strum unintelligently on typewriters for millions of millions of years, would be bound in time to write all the books in the British Museum. If we examined the last page which a particular monkey had typed, and found that it had chanced, in its blind strumming, to type a Shakespeare sonnet, we should rightly regard the occurrence as a remarkable accident, but if we looked through all the millions of pages the monkeys had turned off in untold millions of years, we might be sure of finding a Shakespeare sonnet somewhere amongst them, the product of the blind play of chance. In the same way, millions of millions of stars wandering blindly through space for millions of millions of years are bound to meet with every kind of accident; a limited number are bound to meet with that special kind of accident which calls planetary systems into being. Yet calculation shews that the number of these can at most be very small in comparison with the total number of stars in the sky; planetary systems must be exceedingly rare objects in space.

This rarity of planetary systems is important, because so far as we can see, life of the kind we know on earth could only originate on planets like the earth. It needs suitable physical conditions for its appearance, the most important of which is a temperature at which substances can exist in the liquid state.

The stars themselves are disqualified by being far too hot. We may think of them as a vast collection of fires scattered throughout space, providing warmth in a climate which is at most some four degrees above absolute zero—about 484 degrees of frost on our Fahrenheit scale—and is even lower in the vast stretches of space which lie out beyond the Milky Way. Away from the fires there is this unimaginable cold of hundreds of degrees of frost; close up to them there is a temperature of thousands of degrees, at which all solids melt, all liquids boil.

Life can only exist inside a narrow temperate zone which surrounds each of these fires at a very definite distance.

Outside these zones life would be frozen; inside, it would be shrivelled up. At a rough computation, these zones within which life is possible, all added together, constitute less than a thousand million millionth part of the whole of space. And even inside them, life must be of very rare occurrence, for it is so unusual an accident for suns to throw off planets as our own sun has done, that probably only about one star in 100,000 has a planet revolving round it in the small zone in which life is possible.

Just for this reason it seems incredible that the universe can have been designed primarily to produce life like our own; had it been so, surely we might have expected to find a better proportion between the magnitude of the mechanism and the amount of the product. At first glance at least, life seems to be an utterly unimportant by-product; we living things are somehow off the main line.

We do not know whether suitable physical conditions are sufficient in themselves to produce life. One school of thought holds that as the earth gradually cooled, it was natural, and indeed almost inevitable, that life should come. Another holds that after one accident had brought the earth into being, a second was necessary to produce life. The material constituents of a living body are perfectly ordinary chemical atoms—carbon, such as we find in soot or lampblack; hydrogen and oxygen, such as we find in water; nitrogen, such as forms the greater part of the atmosphere; and so on. Every kind of atom necessary for life must have existed on the new-born earth. At intervals, a group of atoms might happen to arrange themselves in the way in which they are arranged in the living cell. Indeed, given sufficient time, they would be certain to do so, just as certain as the six monkeys would be certain, given sufficient time, to type off a Shakespeare sonnet. But would they then be a living cell? In other words, is a living cell merely a group of ordinary atoms arranged in some non-ordinary way, or is it something more? Is it merely atoms, or is it atoms plus life? Or, to put it in another way, could a sufficiently skilful chemist create life out of the necessary atoms, as a boy can create a machine out of "Meccano," *and then make it go*? We do not know the answer. When it comes it will give us some indication whether other worlds in space are inhabited like ours, and so must have the greatest influence on our interpretation of the meaning of life—it may well produce a greater revolution of thought than Galileo's astronomy or Darwin's biology.

We do, however, know that while living matter consists of quite ordinary atoms, it consists in the main of atoms which have a special capacity for coagulating into extraordinary large bunches or "molecules."

Most atoms do not possess this property. The atoms of hydrogen and oxygen, for instance, may combine to form molecules of hydrogen (H_2 or H_3), of oxygen or ozone (O_2 or O_3), of water (H_2O), or of hydrogen peroxide (H_2O_2), but none of these compounds contains more than four atoms. The addition of nitrogen does not greatly change the situation; the compounds of hydrogen, oxygen and nitrogen all contain comparatively few atoms. But the further addition of carbon completely transforms the picture; the atoms of hydrogen, oxygen, nitrogen *and carbon* combine to form molecules containing hundreds, thousands, and even tens of thousands, of atoms. It is of such molecules that living bodies are mainly formed. Until a century ago it was commonly supposed that some "vital force" was necessary to produce these and the other substances which entered into the composition of the living body. Then Wöhler produced urea ($CO(NH_2)_2$), which is a typical animal product, in his laboratory, by the ordinary processes of chemical synthesis, and other constituents of the living body followed in due course. To-day one phenomenon after another which was at one time attributed to "vital force" is being traced to the action of the ordinary processes of physics and chemistry. Although the problem is still far from solution, it is becoming increasingly likely that what specially distinguishes the matter of living bodies is the presence not of a "vital force," but of the quite commonplace element carbon, always in conjunction with other atoms with which it forms exceptionally large molecules.

If this is so, life exists in the universe only because the carbon atom possesses certain exceptional properties. Perhaps carbon is rather noteworthy chemically as forming a sort of transition between the metals and non-metals, but so far nothing in the physical constitution of the carbon atom is known to account for its very special capacity for binding other atoms together. The carbon atom consists of six electrons revolving around the appropriate central nucleus, like six planets revolving around a central sun; it appears to differ from its two nearest neighbours in the table of chemical elements, the atoms of boron and nitrogen, only in having one electron more than the former and one electron fewer than the latter. Yet this slight difference must account in the last resort for all the difference between life and absence of life. No doubt the reason why the six electron atom possesses these remarkable properties resides somewhere in the ultimate laws of nature, but mathematical physics has not yet fathomed it.

Other similar cases are known to chemistry. Magnetic

phenomena appear in a tremendous degree in iron, and in a lesser degree in its neighbours, nickel and cobalt. The atoms of these elements have 26, 27 and 28 electrons respectively. The magnetic properties of all other atoms are almost negligible in comparison. Somehow, then, although again mathematical physics has not yet unravelled how, magnetism depends on the peculiar properties of the 26, 27 and 28 electron atoms, especially the first. Radio-activity provides a third instance, being confined, with insignificant exceptions, to atoms having from 83 to 92 electrons; again we do not know why.

Thus chemistry can only tell us to place life in the same category as magnetism and radio-activity. The universe is built so as to operate according to certain laws. As a consequence of these laws, atoms having certain definite numbers of electrons, namely 6, 26 to 28, and 83 to 92, have certain special properties, which shew themselves in the phenomena of life, magnetism and radio-activity respectively. An omnipotent creator, subject to no limitations whatever, would not have been restricted to the laws which prevail in the present universe; he might have elected to build the universe to conform to any one of innumerable other sets of laws. If some other set of laws had been chosen, other special atoms might have had other special properties associated with them. We cannot say what, but it seems *à priori* unlikely that either radio-activity or magnetism or *life* would have figured amongst them. Chemistry suggests that, like magnetism and radio-activity, life may merely be an accidental consequence of the special set of laws by which the present universe is governed.

Again the word "accidental" may be challenged. For what if the creator of the universe selected one special set of laws just because they led to the appearance of life? What if this were his way of creating life? So long as we think of the creator as a magnified man-like being, activated by feelings and interests like our own, the challenge cannot be met, except perhaps by the remark that, when such a creator has once been postulated, no argument can add much to what has already been assumed. If, however, we dismiss every trace of anthropomorphism from our minds, there remains no reason for supposing that the present laws were specially selected in order to produce life. They are just as likely, for instance, to have been selected in order to produce magnetism or radio-activity—indeed more likely, since to all appearances physics plays an incomparably greater part in the universe than biology. Viewed from a strictly material standpoint, the utter insignificance of life would seem to go far towards dispelling any idea that it forms a special interest of the Great Architect of the universe.

A trivial analogy may exhibit the situation in a clearer light. An unimaginative sailor, accustomed to tying knots, might think it would be impossible to cross the ocean if tying knots were impossible. Now the capacity for tying knots is limited to space of three dimensions; no knot can be tied in a space of 1, 2, 4, 5 or any other number of dimensions. From this fact our unimaginative sailor might reason that a beneficent creator must have had sailors under his special patronage, and have chosen that space should have three dimensions in order that tying knots and crossing the ocean should be possibilities in the universe he had created—in brief, space was of three dimensions so that there could be sailors. This and the argument outlined above seem to be much on a level, because life as a whole and the tying of knots are pretty much on a level in that neither of them forms more than an utterly insignificant fraction of the total activity of the material universe.

So much for the surprising manner in which, so far as science can at present inform us, we came into being. And our bewilderment is only increased when we attempt to pass from our origins to an understanding of the purpose of our existence, or to foresee the destiny which fate has in store for our race.

Life of the kind we know can only exist under suitable conditions of light and heat; we only exist ourselves because the earth receives exactly the right amount of radiation from the sun; upset the balance in either direction, of excess or defect, and life must disappear from the earth. And the essence of the situation is that the balance is very easily upset.

Primitive man, living in the temperate zone of the earth, must have watched the ice-age descending on his home with something like terror; each year the glaciers came farther down into the valleys; each winter the sun seemed less able to provide the warmth needed for life. To him, as to us, the universe must have seemed hostile to life.

We of these later days, living in the narrow temperate zone surrounding our sun and peering into the far future, see an ice-age of a different kind threatening us. Just as Tantalus, standing in a lake so deep that he only just escaped drowning, was yet destined to die of thirst, so it is the tragedy of our race that it is probably destined to die of cold, while the greater part of the substance of the universe still remains too hot for life to obtain a footing. The sun, having no extraneous supply of heat, must necessarily emit ever less and less of its life-giving radiation, and, as it does so, the temperate zone of space, within which alone life can exist, must close in around it. To remain a possible abode of life, our earth would need to move in ever nearer and nearer to the dying sun. Yet, science tells us that, so far from its moving inwards, inexorable dynamical laws are even now driving it ever farther away from the sun into the outer cold and darkness. And, so far as we can see, they must continue to do so until life is frozen off the earth, unless indeed some celestial collision or cataclysm intervenes to destroy life even earlier by a more speedy death. This prospective fate is not peculiar to our earth; other suns must die like our own, and any life there may be on other planets must meet the same inglorious end.

Physics tells the same story as astronomy. For, independently of all astronomical considerations, the general physical principle known as the second law of thermodynamics predicts that there can be but one end to the universe—a "heat-death" in which the total energy of the universe is uniformly distributed, and all the substance of the universe is at the same temperature. This temperature will be so low as to make life impossible. It matters little by what particular road this final state is reached; all roads lead to Rome, and the end of the journey cannot be other than universal death.

Is this, then, all that life amounts to—to stumble, almost by mistake, into a universe which was clearly not designed for life, and which, to all appearances, is either totally indifferent or definitely hostile to it, to stay clinging on to a fragment of a grain of sand until we are frozen off, to strut our tiny hour on our tiny stage with the knowledge that our aspirations are all doomed to final frustration, and that our achievements must perish with our race, leaving the universe as though we had never been?

Astronomy suggests the question, but it is, I think, mainly to physics that we must turn for an answer. For astronomy can tell us of the present arrangement of the universe, of the vastness and vacuity of space, and of our own insignificance therein; it can even tell us something as to the nature of the changes produced by the passage of time. But we must probe deep into the fundamental nature of things before we can expect to find the answer to our question. And this is not the province of astronomy; rather we shall find that our quest takes us right into the heart of modern physical science.

Chapter II

THE NEW WORLD OF MODERN PHYSICS

Primitive man must have found nature singularly puzzling and intricate. The simplest phenomena could be trusted to recur indefinitely; an unsupported body invariably fell, a stone thrown into water sank, while a piece of wood floated. Yet other more complicated phenomena shewed no such uniformity—the lightning struck one tree in the grove while its neighbour of similar growth and equal size escaped unharmed; one month the new moon brought fair weather, the next month foul.

Confronted with a natural world which was to all appearances as capricious as himself, man's first impulse was to create Nature in his own image; he attributed the seemingly erratic and unordered course of the universe to the whims and passions of gods, or of benevolent or malevolent lesser spirits. Only after much study did the great principle of causation emerge. In time it was found to dominate the whole of inanimate nature: a cause which could be completely isolated in its action was found invariably to produce the same effect. What happened at any instant did not depend on the volitions of extraneous beings, but followed inevitably by inexorable laws from the state of things at the preceding instant. And this state of things had in turn been inevitably determined by an earlier state, and so on indefinitely, so that the whole course of events had been unalterably determined by the state in which the world found itself at the first instant of its history; once this had been fixed, nature could move only along one road to a predestined end. In brief, the act of creation had created not only the universe but its whole future history. Man, it is true, still believed that he himself was able to affect the course of events by his own volition, although in this he was guided by instinct rather than by logic, science, or experience, but henceforth the law of causation took charge of all such events as he had previously assigned to the actions of supernatural beings.

The final establishment of this law as the primary guiding principle in nature was the triumph of the seventeenth century, the great century of Galileo and Newton. Apparitions in the sky were shewn to result merely from the universal laws of optics; comets, which had hitherto been regarded as portents of the fall of empires or the death of kings, were proved to have their motions prescribed by the universal law of gravitation. "And," wrote Newton, "would that the rest of the phenomena of nature could be deduced by a like kind of reasoning from mechanical principles."

Out of this resulted a movement to interpret the whole material universe as a machine, a movement which steadily gained force until its culmination in the latter half of the nineteenth century. It was then that Helmholtz declared that "the final aim of all natural science is to resolve itself into mechanics," and Lord Kelvin confessed that he could understand nothing of which he could not make a mechanical model. He, like many of the great scientists of the nineteenth century, stood high in the engineering profession; many others could have done so had they tried. It was the age of the engineer-scientist, whose primary ambition was to make mechanical models of the whole of nature. Waterston, Maxwell and others had explained the properties of a gas as machine-like properties with great success; the machine consisted of a vast multitude of tiny round, smooth spheres, harder than the hardest steel, flying about like a hail of bullets on a battlefield. The pressure of a gas, for instance, was caused by the impact of the speedily flying bullets; it was like the pressure which a hailstorm exerts on the roof of a tent. When sound was transmitted through a gas, these bullets were the messengers. Similar attempts were made to explain the properties of liquids and solids as machine-like properties, although with considerably less success, and also on light and gravitation—with no success at all. Yet this want of success failed to shake the belief that the universe must in the last resort admit of a purely mechanical interpretation. It was felt that only greater efforts were needed, and the whole of inanimate nature would at last stand revealed as a perfectly acting machine.

All this had an obvious bearing on the interpretation of human life. Each extension of the law of causation, and each success of the mechanical interpretation of nature, made the belief in free-will more difficult. For if all nature obeyed the law of causation, why should life be exempt? Out of such considerations arose the mechanistic philosophies of the seventeenth and eighteenth centuries, and their natural reactions, the idealist philosophies which succeeded them. Science appeared to favour a mechanistic view which saw the whole material world as a vast machine. By contrast, the idealistic view (p. 125 below) attempted to regard the world as the creation of thought and so as consisting of thought.

Until early in the nineteenth century it was still compatible with scientific knowledge to regard life as something standing entirely apart from inanimate nature. Then came the discovery that living cells were formed of precisely the same chemical atoms as non-living matter, and so were presumably governed by the same natural laws. This led to the question why the particular atoms of which our bodies and brains were formed should be exempt from the laws of causation. It began to be not only conjectured, but even fiercely maintained, that life itself must, in the last resort, prove to be purely mechanical in its nature. The mind of a Newton, a Bach or a Michelangelo, it was said, differed only in complexity from a printing press, a whistle or a steam saw; their whole function was to respond exactly to the stimuli they received from without. Because such a creed left no room for the operation of choice and free-will, it removed all basis for morality. Paul did not choose to be different from Saul; he could not help being different; he was affected by a different set of external stimuli.

An almost kaleidoscopic re-arrangement of scientific thought came with the change of century. The early scientists were only able to study matter in chunks large enough to be directly apprehended by the unaided senses; the tiniest piece of matter with which they could experiment contained millions of millions of molecules. Pieces of this size undoubtedly behaved in a mechanical way, but this provided no guarantee that single molecules would behave in the same way; everyone knows the vast difference between the behaviour of a crowd and that of the individuals that compose it.

At the end of the nineteenth century it first became possible to study the behaviour of single molecules, atoms and electrons. The century had lasted just long enough for science to discover that certain phenomena, radiation and gravitation in particular, defied all attempts at a purely mechanical explanation. While philosophers were still debating whether a machine could be constructed to reproduce the thoughts of Newton, the emotions of Bach or the inspiration of Michelangelo, the average man of science was rapidly becoming convinced that no machine could be constructed to reproduce the light of a candle or the fall of an apple. Then, in the closing months of the century, Professor Max Planck of Berlin brought forward a tentative explanation of certain phenomena of radiation which had so far completely defied interpretation. Not only was his explanation non-mechanical in its nature; it seemed impossible to connect it up with any mechanical line of thought. Largely for this reason, it was criticised, attacked and even ridiculed. But it proved brilliantly successful, and ultimately developed into the modern "quantum theory," which forms one of the great dominating principles of modern physics. Also, although this was not apparent at the time, it marked the end of the mechanical age in science, and the opening of a new era.

In its earliest form, Planck's theory hardly went beyond suggesting that the course of nature proceeded by tiny jumps and jerks, like the hands of a clock. Yet, although it does not advance continuously, a clock is purely mechanical in its ultimate nature, and follows the law of causation absolutely. Einstein shewed in 1917 that the theory founded by Planck appeared, at first sight at least, to entail consequences far more revolutionary than mere discontinuity. It appeared to dethrone the law of causation from the position it had heretofore held as guiding the course of the natural world. The old science had confidently proclaimed that nature could follow only one road, the road which was mapped out from the beginning of time to its end by the continuous chain of cause and effect; state A was inevitably succeeded by state B. So far the new science has only been able to say that state A may be followed by state B or C or D or by innumerable other states. It can, it is true, say that B is more likely than C, C than D, and so on; it can even specify the relative probabilities of states B, C and D. But, just because it has to speak in terms of probabilities, it cannot predict with certainty which state will follow which; this is a matter which lies on the knees of the gods—whatever gods there be.

A concrete example will explain this more clearly. It is known that the atoms of radium, and of other radio-active substances, disintegrate into atoms of lead and helium with the mere passage of time, so that a mass of radium continually diminishes in amount, being replaced by lead and helium. The law which governs the rate of diminution is very remarkable. The amount of radium decreases in precisely the same way as a population would if there were no births, and a uniform death-rate which was the same for every individual, *regardless of his age*. Or again, it decreases in the same way as the numbers of a battalion of soldiers who are exposed to absolutely random undirected fire. In brief, old age appears to mean nothing to the individual radium atom; it does not die because it has lived its life, but rather because in some way fate knocks at the door.

To take a concrete illustration, suppose that our room contains two thousand atoms of radium. Science cannot say how many of these will survive after a year's time, it can only tell us the relative odds in favour of the number being 2000, 1999, 1998, and so on. Actually the most likely event is that the number will be 1999; the probabilities are in favour of one, and only one, of the 2000 atoms breaking up within the next year.

We do not know in what way this particular atom is selected out of the 2000. We may at first feel tempted to conjecture it will be the atom that gets knocked about most or gets into the hottest places, or what not, in the coming year. Yet this cannot be, for if blows or heat could disintegrate one atom, they could disintegrate the other 1999, and we should be able to expedite the disintegration of radium merely by compressing it or heating it up. Every physicist believes this to be impossible; he rather believes that every year fate knocks at the door of one radium atom in every 2000, and compels it to break up; this is the hypothesis of "spontaneous disintegration" advanced by Rutherford and Soddy in 1903.

History of course may repeat itself, and once again an apparent capriciousness in nature may be found, in the light of fuller knowledge, to arise out of the inevitable operation of the law of cause and effect. When we speak in terms of probabilities in ordinary life, we merely shew that our knowledge is incomplete; we may say it appears probable that it will rain to-morrow, while the meteorological expert, knowing that a deep depression is coming eastward from the Atlantic, can say with confidence that it will be wet. We may speak of the odds on a horse, while the owner knows it has broken its leg. In the same way, the appeal of the new physics to probabilities may merely cloak its ignorance of the true mechanism of nature.

An illustration will suggest how this might be. Early in the present century, McLennan, Rutherford and others detected in the earth's atmosphere a new type of radiation, distinguished by its extremely high powers of penetrating solid matter. Ordinary light will penetrate only a fraction of an inch through opaque matter; we can shield our faces from the rays of the sun with a sheet of paper, or an even thinner screen of metal. The X-rays have a far greater penetrating power; they can be made to pass through our hands, or even our whole bodies, so that the surgeon can photograph our bones. Yet metal of the thickness of a coin stops them completely. But the radiation discovered by McLennan and Rutherford could penetrate through several yards of lead or other dense metal.

We now know that a large part of this radiation, generally described as "cosmic radiation," has its origin in outer space. It falls on the earth in large quantities, and its powers of destruction are immense. Every second it breaks up about twenty atoms in every cubic inch of our atmosphere, and millions of atoms in each of our bodies. It has been suggested that this radiation, falling on germ-plasm, may produce the spasmodic biological variations which the modern theory of evolution demands; it may have been cosmic radiation that turned monkeys into men.

In the same way, it was at one time conjectured that the falling of cosmic radiation on radio-active atoms might be the cause of their disintegration. The rays fell like fate, striking now one atom and now another, so that the atoms succumbed like soldiers exposed to random fire, and the law which governed their rate of disappearance was explained. This conjecture was disproved by the simple device of taking radio-active matter down a coal-mine. It was now completely shielded from the cosmic rays, but continued to disintegrate at the same rate as before.

This hypothesis failed, but probably many physicists expect that some other physical agency may yet be found to act the rôle of fate in radio-active disintegration. The death-rate of atoms would obviously then be proportional to the strength of this agency. But other similar phenomena present far greater difficulties.

Amongst these is the familiar phenomenon of the emission of light by an ordinary electric-light bulb. The essentials are that a hot filament receives energy from a dynamo and discharges it as radiation. Inside the filament, the electrons of millions of atoms are whirling round in their orbits, every now and then jumping, suddenly and almost discontinuously, from one orbit to another, sometimes emitting, and sometimes absorbing, radiation in the process. In 1917, Einstein investigated what may be described as the statistics of these jumps. Some are of course caused by the radiation itself and the heat of the filament. But these are not enough to account for the whole of the radiation emitted by the filament. Einstein found that there must be other jumps as well, and that these must occur spontaneously, like the disintegration of the radium atom. In brief, it appears as though fate must be invoked here also. Now if some ordinary physical agency played the part of fate in this case, its strength ought to affect the intensity of the emission of radiation by the filament. But, so far as we know, the intensity of the radiation depends only on known constants of nature, which are the same here as in the remotest stars. And this seems to leave no room for the intervention of an external agency.

We can perhaps form some sort of a picture of the nature of these spontaneous disintegrations or jumps, by comparing the atom to a party of four card-players who agree to break up as soon as a hand is dealt in which each player receives just one complete suit. A room containing millions of such parties may be taken to represent a mass of radio-active substance. Then it can be shewn that the number of card parties will decrease according to the exact law of radio-active decay on one condition—*that the cards are well shuffled between each deal.* If there is adequate shuffling of the cards, the passage of time and the past will mean nothing to the card players, for the situation is born afresh each time the cards are shuffled. Thus the death-rate per thousand will be constant, as with atoms of radium. But if the cards are merely taken up after each deal, without shuffling, each deal follows inevitably from the preceding, and we have the analogue of the old law of causation. Here the rate of diminution in the number of players would be different from that actually observed in radio-active disintegration. We can only reproduce this by supposing the cards to be continually shuffled, and the shuffler is he whom we have called fate.

Thus, although we are still far from any positive knowledge, it seems possible that there may be some factor, for which we have so far found no better name than fate, operating in nature to neutralise the cast-iron inevitability of the old law of causation. The future may not be as unalterably determined by the past as we used to think; in part at least it may rest on the knees of whatever gods there be.

Many other considerations point in the same direction. For instance, Professor Heisenberg has shewn that the concepts of the modern quantum theory involve what he calls a "principle of indeterminacy." We have long thought of the workings of nature as exemplifying the acme of precision. Our man-made machines are, we know, imperfect and inaccurate, but we have cherished a belief that the innermost workings of the atom would exemplify absolute accuracy and precision. Yet Heisenberg now makes it appear that nature abhors accuracy and precision above all things.

According to the old science, the state of a particle, such as an electron, was completely specified when we knew its position in space at a single instant and its speed of motion through space at the same instant. These data, together with a knowledge of any forces which might act on it from outside, determined the whole future of the electron. If these data were given for all the particles in the universe, the whole future of the universe could be predicted.

The new science, as interpreted by Heisenberg, asserts that these data are, from the nature of things, unprocurable. If we know that an electron is at a certain point in space, we cannot specify exactly the speed with which it is moving—nature permits a certain "margin of error," and if we try to get within this margin, nature will give us no help: she knows nothing, apparently, of absolutely exact measurements. In the same way, if we know the exact speed of motion of an electron, nature refuses to let us discover its exact position in space. It is as though the position and motion of the electron had been marked on the two different faces of a lantern slide. If we put the slide in a bad lantern, we can focus half-way between the two faces, and shall see both the position and motion of the electron tolerably clearly. With a perfect lantern, we could not do this; the more we focussed on one, the more blurred the other would become.

The imperfect lantern is the old science. It gave us the illusion that, if only we had a perfect lantern, we should be able to determine both the position and motion of a particle at a given instant with perfect sharpness, and it was this illusion that introduced determinism into science. But now that we have the more perfect lantern in the new science, it merely shews us that the specifications of position and motion lie in two different planes of reality, which cannot be brought simultaneously into sharp focus. In so doing, it cuts away the ground on which the old determinism was based.

Or again, to take another analogy, it is almost as though the joints of the universe had somehow worked loose, as though its mechanism had developed a certain amount of "play," such as we find in a well-worn engine. Yet the analogy is misleading if it suggests that the universe is in any way worn out or imperfect. In an old or worn engine, the degree of "play" or "loose jointedness" varies from point to point; in the natural world it is measured by the mysterious quantity known as "Planck's constant *h*," which proves to be absolutely uniform throughout the universe. Its value, both in the laboratory and in the stars, can be measured in innumerable ways, and always proves to be precisely the same. Yet the fact that "loose jointedness," of any type whatever, pervades the whole universe destroys the case for absolutely strict causation, this latter being the characteristic of perfectly fitting machinery.

The uncertainty to which Heisenberg has called attention is partially, but not wholly, of a subjective nature. The fact that we cannot specify the position and speed of an electron with absolute precision arises in part from the clumsiness of the apparatus with which we work—just as a man cannot weigh himself with absolute accuracy if he has no weight less than a pound at his disposal. The smallest unit known to science is an electron, so that no smaller unit can possibly be at the disposal of the physicist. In actual fact, it is not the finite size of this unit that is the immediate cause of the trouble, so much as that of the mysterious unit *h* introduced by Planck's quantum theory. This measures the size of the "jerks" by which nature moves, and so long as these jerks are of finite size, it is as impossible to make exact measurements as to weigh oneself exactly on a balance which can only move by jerks.

This subjective uncertainty has, however, no bearing on the problems of radio-activity and radiation discussed on pp. 18 and 21. And there are many other phenomena of nature, too numerous even to enumerate here, which cannot be included in any consistent scheme unless the conception of indeterminacy is introduced somewhere and somehow.

These and other considerations to which we shall return below (pp. 34, 107) have led many physicists to suppose that there is no determinism in events in which atoms and electrons are involved singly, and that the apparent determinism in large-scale events is only of a statistical nature. Dirac describes the situation as follows:

> When an observation is made on any atomic system...in a given state, the result will not in general be determinate, *i.e.* if the experiment is repeated several times under identical conditions, several different results may be obtained. If the experiment is repeated a large number of times, it will be found that each particular result will be obtained a definite fraction of the total number of times, so that one can say there is a definite probability of its being obtained any time the experiment is performed. This probability the theory enables one to calculate. In special cases, the probability may be unity, and the result of the experiment is then quite determinate.

In other words, when we are dealing with atoms and

electrons in crowds, the mathematical law of averages imposes the determinism which physical laws have failed to provide.

We can illustrate the concept by an analogous situation in the large-scale world. If we spin a half-penny, nothing within our knowledge may be able to decide whether it will come down heads or tails, yet if we throw up a million tons of half-pence, we know there will be 500,000 tons of heads and 500,000 tons of tails. The experiment may be repeated time after time, and will always give the same result. We may be tempted to instance it as evidence of the uniformity of nature, and to infer the action of an underlying law of causation: in actual fact it is an instance only of the operation of the purely mathematical laws of chance.

Yet the number of half-pence in a million tons is nothing in comparison with the number of atoms in even the smallest piece of matter with which the earlier physicists could experiment. It is easy to see how the illusion of determinacy—if it is an illusion—crept into science.

We have still no definite knowledge on any of these problems. A number, although I think a rapidly diminishing number, of physicists still expect that in some way the law of strict causation will in the end be restored to its old place in the natural world, but the recent trend of scientific progress gives them no encouragement. At any rate, the concept of strict causation finds no place in the picture of the universe which the new physics presents to us, with the result that this picture contains more room than did the old mechanical picture for life and consciousness to exist within the picture itself, together with the attributes which we commonly associate with them, such as free-will, and the capacity to make the universe in some small degree different by our presence. For, for aught we know, or for aught that the new science can say to the contrary, the gods which play the part of fate to the atoms of our brains may be our own minds. Through these atoms our minds may perchance affect the motions of our bodies and so the state of the world around us. To-day science can no longer shut the door on this possibility; she has no longer any unanswerable arguments to bring against our innate conviction of free-will. On the other hand, she gives no hint as to what absence of determinism or causation may mean. If we, and nature in general, do not respond in a unique way to external stimuli, what determines the course of events? If anything at all, we are thrown back on determinism and causation; if nothing at all, how can anything ever occur?

As I see it, we are unlikely to reach any definite conclusions on these questions until we have a better understanding of the true nature of time. The fundamental laws of nature, in so far as we are at present acquainted with them, give no reason why time should flow steadily on: they are equally prepared to consider the possibility of time standing still or flowing backwards. The steady onward flow of time, which is the essence of the cause-effect relation, is something which we superpose on to the ascertained laws of nature out of our own experience; whether or not it is inherent in the nature of time, we simply do not know, although, as we shall see shortly, the theory of relativity goes at any rate some distance towards stigmatising this steady onward flow of time and the cause-effect relation as illusions; it regards time merely as a fourth dimension to be added to the three dimensions of space, so that *post hoc ergo propter hoc* may be no more true of a sequence of happenings in time than it is of the sequence of telegraph-poles along the Great North Road.

It is always the puzzle of the nature of time that brings our thoughts to a standstill. And if time is so fundamental that an understanding of its true nature is for ever beyond our reach, then so also in all probability is a decision in the age-long controversy between determinism and free-will.

The possible abolition of determinism and the law of causation from physics are, however, comparatively recent developments in the history of the quantum theory. The primary object of the theory was to explain certain phenomena of radiation, and to understand the question at issue we must retrace our steps as far back as Newton and the seventeenth century.

The most obvious fact about a ray of light, at any rate to superficial observation, is its tendency to travel in a straight line; everyone is familiar with the straight edges of a sunbeam in a dusty room. As a rapidly moving particle of matter also tends to travel in a straight line, the early scientists, rather naturally, thought of light as a stream of particles thrown out from a luminous source, like shot from a gun. Newton adopted this view, and added precision to it in his "corpuscular theory of light."

Yet it is a matter of common observation that a ray of light does not always travel in a straight line. It can be abruptly turned by reflection, such as occurs when it falls on the surface of a mirror. Or its path may be bent by refraction, such as occurs when it enters water or any liquid medium; it is refraction that makes our oar look broken at the point where it enters the water, and makes the river look shallower than it proves to be when we step into it. Even in Newton's time the laws which governed these phenomena were well known. In the case of reflection the angle at which the ray of light struck the mirror was exactly the same as that at which it came off after reflection; in other words, light bounces off a mirror like a tennis-ball bouncing off a perfectly hard tennis-court. In the case of refraction, the sine of the angle of incidence stood in a constant ratio to the sine of the angle of refraction. We find Newton at pains to shew that his light-corpuscles would move in accordance with these laws, if they were subjected to certain definite forces at the surfaces of a mirror or a refracting liquid. Here are Propositions XCIV and XCVI of the *Principia*:

PROPOSITION XCIV

If two similar mediums be separated from each other by a space terminated on both sides by parallel planes, and a body in its passage through that space be attracted or impelled perpendicularly towards either of those mediums, and not agitated or hindered by any other force; and the attraction be every where the same at equal distances from either plane, taken towards the same hand of the plane; I say, that the sine of incidence upon either plane will be to the sine of emergence from the other plane in a given ratio.

PROPOSITION XCVI

The same things being supposed, and that the motion before incidence is swifter than afterwards; I say, that if the line of incidence be inclined continually, the body will be at last reflected, and the angle of reflexion will be equal to the angle of incidence.

Newton's corpuscular theory met its doom in the fact that when a ray of light falls on the surface of water, only part of it is refracted. The remainder is reflected, and it is this latter part that produces the ordinary reflections of objects in a lake, or the ripple of moonlight on the sea. It was objected that Newton's theory failed to account for this reflection, for if light had consisted of corpuscles, the forces at the surface of the water ought to have treated all corpuscles alike; when one corpuscle was refracted all ought to be, and this left water with no power to reflect the sun, moon or stars. Newton tried to obviate this objection by attributing "alternate fits of transmission and reflection" to the surface of the water—the corpuscle which fell on the surface at one instant was admitted, but the next instant the gates were shut, and its companion was turned away to form reflected light. This concept was strangely and strikingly anticipatory of modern quantum theory in its abandonment of the uniformity of nature and its replacement of determinism by probabilities, but it failed to carry conviction at the time.

And, in any case, the corpuscular theory was confronted by other and graver difficulties. When studied in sufficiently minute detail, light is not found to travel in such absolutely straight lines as to suggest the motions of particles. A big object, such as a house or a mountain, throws a definite shadow, and so gives as good protection from the glare of the sun as it would from a shower of bullets. But a tiny object, such as a very thin wire, hair or fibre, throws no such shadow. When we hold it in front of a screen, no part of the screen remains unilluminated. In some way the light contrives to bend round it, and, instead of a definite shadow, we see an alternation of light and comparatively dark parallel bands, known as "interference bands." To take another instance, a large circular hole in a screen lets through a circular patch of light. But make the hole as small as the smallest of pinholes, and the pattern thrown on a screen beyond is not a tiny circular patch of light, but a far larger pattern of concentric rings, in which light and dark rings alternate—"diffraction rings." Fig. 1 of Plate II (p. 37) shews the pattern obtained by allowing a beam of light to pass through a pinhole on to a photographic plate. All the light which is more than a pinhole's radius from the centre has in some way bent round the edge of the hole.

Newton regarded these phenomena as evidence that his "light-corpuscles" were attracted by solid matter. He wrote:

> The rays of light that are in our air, in their passage near the angles of bodies, whether transparent or opaque (such as the circular and rectangular edges of coins, or of knives, or broken pieces of stone or glass), are bent or inflected round those bodies, as if they were attracted to them; and those rays which in their passage came nearest to the bodies are the most inflected, as if they were most attracted.

Here again Newton was strangely anticipatory of present-day science, his supposed forces being closely analogous to the "quantum forces" of the modern wave-mechanics. But they failed to give any detailed explanation of diffraction-phenomena, and so met with no favour.

In time, all these and similar phenomena were adequately explained by supposing that light consists of waves, somewhat similar to those which the wind blows up on the sea, except that, instead of each wave being many yards long, many thousands of waves go to a single inch. Waves of light bend round a small obstacle in exactly the way in which waves of the sea bend round a small rock. A rocky reef miles long gives almost perfect shelter from the sea, but a small rock gives no such protection—the waves pass round it on either side, and re-unite behind it, just as waves of light re-unite behind our thin hair or fibre. In the same way sea-waves which fall on the entrance to a harbour do not travel in a straight line across the harbour but bend round the edges of the breakwater, and make the whole surface of the water in the harbour rough. Fig. 1 of Plate II (p. 87) shews the "roughness" beyond a pinhole produced by waves of light which have bent round the edges of the pinhole like sea-waves bending round a breakwater. The seventeenth century regarded light as a shower of particles, the eighteenth century, discovering that this was inadequate to account for small-scale phenomena such as we have just described, replaced the showers of particles by trains of waves.

Yet the replacement brought its own difficulties with it. When sunlight is passed through a prism, it is broken up into a rainbow-like "spectrum" of colours—red, orange, yellow, green, blue, indigo and violet. If light consisted of waves like the waves of the sea, it can be shewn that all the light of the analysed sunlight ought to be found at the extreme violet end of the spectrum. Not only so, but extreme violet waves have an unlimited capacity for absorbing energy, and as they have their mouths permanently wide open, all the energy of the universe would rapidly pass into the form of violet, or ultra-violet, radiation travelling through space.

The "quantum theory" came into being as an effort to cure the wave theory of light of these defects. It has been completely successful. It has shewn that Newton was not wholly wrong in regarding light as corpuscular, for it has proved that a beam of light may be regarded as broken up into discrete units, called "light-quanta" or "photons," with almost the definiteness with which a shower of rain may be broken up into drops of water, a shower of bullets into separate pieces of lead, or a gas into separate molecules.

At the same time, the light does not lose its undulatory character. Each little parcel of light has a definite quantity, of the nature of a length, associated with it. We call this its "wave-length," because when the light in question is passed through a prism, it behaves exactly as waves of this particular length of wave would do. Light of long wave-length is made up of small parcels, and *vice-versa,* the amount of energy in each parcel being inversely proportional to this wave-length, so that we can always calculate the energy of a photon from its wave-length, and *vice-versa.*

It is impossible even to summarise the great mass of evidence on which these concepts are based. It all, absolutely without exception, indicates that light travels through laboratory apparatus in unbroken photons; no observation yet made has revealed the existence of a fraction of a photon, or given any reason for suspecting that such a thing can exist. Two examples may typify the whole.

Radiation may, under suitable conditions, break up the atoms on which it falls. A study of the shattered atoms discloses how much energy has been let loose on each to do the work of breaking it up. Invariably the energy proves to be exactly that of a complete photon, as calculated from its known wave-length. It is as though an army of light had come into conflict with an army of matter. It has long been known that the latter army consists of individual soldiers, the atoms; it now appears that the former also consists of individual soldiers, the photons, a study of the battlefield shewing that the conflict has consisted of individual man-to-man encounters.

As a second example, Professor Compton of Chicago has recently studied what happens when X-radiation falls on electrons. He finds that the radiation is scattered exactly as though it consisted of material particles of light, or photons, moving as separate detached units, this time like bullets on a battlefield, and hitting all electrons which stand in their way. The extent to which individual photons are deflected from their courses at these collisions makes it possible to calculate the energy of the photons, and again this is found to agree exactly with that calculated from their wave-length.

This concept of indivisible photons again leads us back to indeterminacy. There are various methods of splitting up a beam of light into two parts which follow different paths. When the beam is reduced to a single photon, it must follow either one path or the other; it cannot distribute itself over both because the photon is indivisible. And its choice of path proves to be a matter of probability, not of determinacy.

In this way it appears that the seventeenth century, which regarded light as mere particles, and the nineteenth century, which regarded it as mere waves, were both wrong —or, if we prefer, both right. Light, and indeed radiation of all kinds, is both particles and waves at the same time. In Professor Compton's experiments, X-radiation falls on single electrons and behaves like a shower of discrete particles; in the experiments of Laue, Bragg and others, exactly similar radiation falls on a solid crystal and behaves in all respects like a succession of waves. And it is the same throughout nature; the same radiation can simulate both particles and waves at the same time. Now it behaves like particles, now like waves; no general principle yet known can tell us what behaviour it will choose in any particular instance.

Clearly we can only preserve our belief in the uniformity of nature by making the supposition that particles and waves are in essence the same thing. And this brings us to the second, and far more exciting, half of our story. The first half, which has just been told, is that radiation can appear now as waves and now as particles; the second is that electrons and protons, the fundamental units of which all matter is composed (p. 46), can also appear now as particles, and now as waves. A duality has recently been discovered in the nature of electrons and protons similar to that already known to exist in the nature of radiation; these also appear to be particles and waves at the same time.

When Newton's corpuscular theory of light first gave place to the undulatory theory, it became necessary to explain how a succession of waves could simulate the behaviour of a shower of particles, and move in a straight line except where it was deflected from its course by reflection or refraction. For if the sunbeam let in through a crack in the shutter consisted of waves, it was natural to expect that they would spread through the whole of the room, just as a ripple spreads over the whole surface of a pond, or as the very narrow beam which has passed through a pinhole has spread out in Fig. 1 of Plate II (p. 37). Yet Young and Fresnel shewed that an undisturbed succession of waves of sufficient width would move as a beam, without appreciable sideways spread—like a shower of freely moving particles—and would be reflected from a mirror in the same way in which a projectile bounces off a perfectly hard surface. It was also shewn that such a system of waves would be refracted according to the known laws of refraction of light. Finally, if such a system of waves travelled through a medium whose refracting power changed continuously, its path would be similar to that of a particle which was made to deviate from a straight path by continuously acting forces. Indeed the two paths could be made identical by taking the force at every point proportional to the change in the square of the refractive index. This explained the success of Newton's Propositions XCIV and XCVI which we have quoted on p. 29.

Thus whatever the particles of Newton's corpuscular theory could do, a succession of waves could do the same. But, just because of their greater complexity, they were able to do more, and in every case in which the particles failed to simulate the behaviour of light, it was found that a system of waves could fill the part completely. In this way Newton's supposed particles became resolved into systems of waves.

The last few years have seen the particles of which ordinary matter is formed—i.e. protons and electrons—resolved into systems of waves in a somewhat similar way. In many circumstances, the behaviour of an electron or proton is found to be too complex to permit of explanation as the motion of a mere particle; Louis de Broglie, Schrödinger and others have accordingly tried to interpret it as the behaviour of a group of waves and, in so doing, have founded the branch of mathematical physics which is now known as "Wave-Mechanics."

If we watch an ordinary tennis-ball bouncing off the surface of a perfectly hard tennis-court, we shall find that its motion is the same as that of a beam of light reflected at the surface of a mirror, so that we may properly speak of the ball as being "reflected" from the surface of the court. But

PLATE II. THE DIFFRACTION OF LIGHT AND OF ELECTRONS

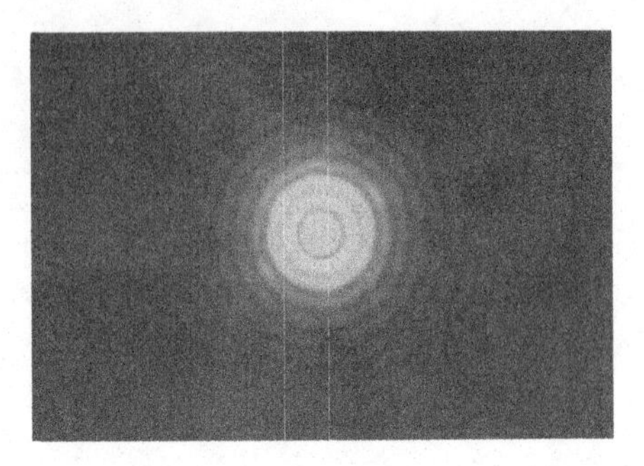

Fig. 1. Diffraction rings produced by light passing through a minute aperture (a pinhole) in an opaque screen (L. R. Wilberforce)

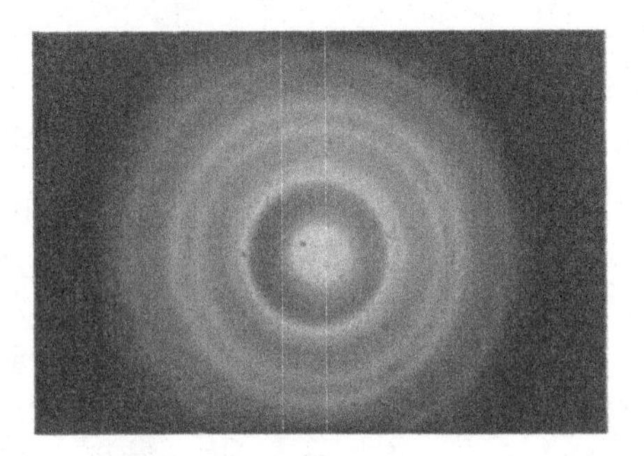

Fig. 2. Diffraction rings produced by electrons passing through a minute area of gold film (G. P. Thomson)

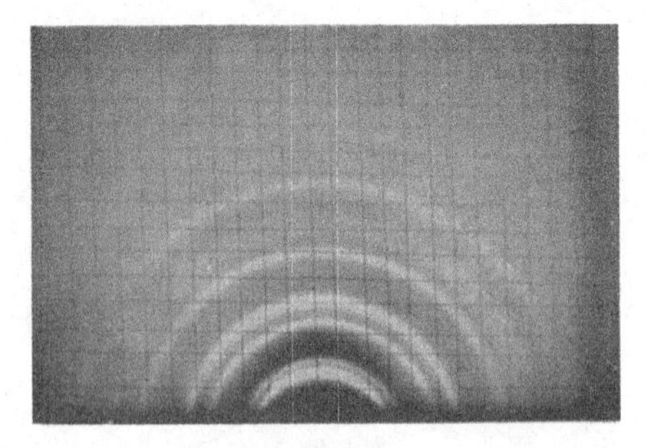

Fig. 3. Diffraction rings produced by electrons reflected off a minute area of a gold surface (G. P. Thomson)

there is not much gained by the discovery. No doubt it would permit us to interpret a tennis-ball as a system of waves if we desired to do so, but we do not; for one thing we can see, or think we can see, that a tennis-ball is not a system of waves.

The case would be different if the moving object were not a tennis-ball but an electron. If the motion of an electron bouncing off a surface were observed to be like that of a system of waves, nothing could preclude the possibility of the electron being a system of waves. No one can now say "This does not interest me—I can see the electron, and it clearly is not a system of waves," for no one has ever seen an electron, or has the remotest conception as to what it would look like. We are just as free *à priori* to consider an electron as a system of waves, as to consider Newton's light-corpuscles as systems of waves. And to find out whether an electron really is a system of waves, we must turn to phenomena in which a hard particle and a system of waves would behave differently.

Now the phenomena in which the electron did not behave at all as it was expected to, so long as it was regarded as a particle, provide precisely the group of phenomena we want, and in every case the electron is found to behave exactly like a system of waves. One particular phenomenon is that of a shower of electrons bouncing off a metal plate; they do not bounce off like a shower of hailstones or tennis-balls, but produce a diffraction pattern (p. 31) as a system of waves would do (see Plate II, Fig. 3). And it is the same when the shower of electrons is shot through a tiny aperture; they spread laterally and produce a diffraction pattern very similar to that produced by waves of light (see Plate II, Figs. 1 and 2). This does not of course prove that an electron actually consists of waves, but it raises the question whether a system of waves does not provide a better picture of the electron than the hard particle. Actually a system of waves provides a picture which has never yet failed to predict the behaviour of the electron, while the conception of an electron as a hard particle has failed on innumerable occasions.

The new wave-mechanics shews that a moving electron or proton ought to behave like a system of waves of quite definite wave-length; this depends on the mass of the moving particle, and on its speed of motion, but on nothing else. And the wave-lengths it assigns to electrons and protons moving under ordinary laboratory conditions are such as can be easily measured with ordinary laboratory apparatus.

Experiments on what may properly be described as the reflection and refraction of electrons have been performed by Davisson and Germer in America, by Professor G. P. Thomson at Aberdeen, by Rupp in Germany, by Kikuchi in Japan, and by many others. Moving electrons are shot, as a parallel beam, either on to or through a metallic surface. And in each case the effect recorded on a suitably placed photographic plate is not at all that which would be observed if the electrons behaved like a shower of small shot or other hard particles. A diffraction pattern is invariably obtained, consisting of a system of concentric rings, light and dark rings alternating. The pattern is the same as would have been produced if waves of a certain definite wave-length had fallen on the metal, and when the wave-length is measured it proves to be exactly that predicted by the wave-mechanics formula already mentioned.

Recently Professor A. J. Dempster of Chicago has had a similar success with moving protons.

These and other experiments make it clear that the waves and wave-lengths associated with moving electrons and protons are at least something more than a pure myth. Something of an undulatory nature is certainly involved, and the picture which represents moving electrons and protons as systems of waves explains their behaviour far better, both inside and outside atoms, than did the old picture which regarded them merely as charged particles.

We shall discuss the nature of these waves more fully below (p. 107). For our immediate purpose it is enough that the ingredients of matter (electrons and protons) and radiation both exhibit a dual nature. So long as science deals only with large-scale phenomena, an adequate picture can generally be obtained by supposing both to be of the nature of particles. But when science comes to closer grips with nature, and passes to the study of small-scale phenomena, matter and radiation are found equally to resolve themselves into waves.

If we want to understand the fundamental nature of the physical universe, it is to these small-scale phenomena that we must turn our attention. Here the ultimate nature of things lies hidden, and what we are finding is waves.

In this way, we are beginning to suspect that we live in a universe of waves, and nothing but waves. We shall discuss the nature of these waves below. At the moment it is enough to notice that modern science has travelled very far from the old view which regarded the universe merely as a collection of hard bits of matter in which waves of radiation occasionally appeared as an incident. And the next chapter will carry us farther along the same road.

Chapter III

MATTER AND RADIATION

In the early days of science, the unquestioning acceptance of the law of causation as a guiding principle in the natural world led to the discovery and formulation of laws of the general type "an assigned cause *A* leads to a known effect *B*." For instance the addition of heat to ice causes it to melt, or stated in more detail, heat decreases the amount of ice in the universe and increases the amount of water.

Primitive man would become acquainted with this law very easily—he had only to watch the action of the sun on hoar-frost, or the effect of the long summer days on the mountain glaciers. In winter he would notice that cold changed water back into ice. At a farther stage it might be discovered that the re-frozen ice was equal in amount to the original ice before melting. It would then be a natural inference that something belonging to a more general category than either water or ice had remained unaffected in amount throughout the transformation

$$\text{ice} \rightarrow \text{water} \rightarrow \text{ice}.$$

Modern physics is familiar with laws of this type, which it describes as "conservation laws." The discovery we have just attributed to primitive man is a special case of the law of conservation of matter. The law of "conservation of X," whatever X may be, means that the total amount of X in the universe remains perpetually the same: nothing can change X into something which is not X. Every such law is of necessity hypothetical; what it actually expresses is that nothing we have so far done has succeeded in changing the total amount of X. And if we have tried enough things and failed every time, it is legitimate to propound a law of conservation of X, at any rate as a working hypothesis.

At the end of last century, physical science recognised three major conservation laws:

A the conservation of matter,
B " " mass,
C " " energy.

Other minor laws, such as those of the conservation of linear and angular momenta, need not enter our discussion, since they are mere deductions from the three major laws already mentioned.

Of the three major laws, the conservation of matter was the most venerable. It had been implied in the atomistic philosophy of Democritus and Lucretius, which supposed all matter to be made up of uncreatable, unalterable and indestructible atoms. It asserted that the matter content of the universe remained always the same, and the matter content of any bit of the universe or of any region of space remained the same except in so far as it was altered by the ingress or egress of atoms. The universe was a stage in which always the same actors—the atoms—played their parts, differing in disguises and groupings, but without change of identity. And these actors were endowed with immortality.

The second law, that of the conservation of mass, was of more modern growth. Newton had supposed every body or piece of substance to have associated with it an unvarying quantity, its mass, which gave a measure of its "inertia" or reluctance to change its motion. If one motor-car requires twice the engine power of another to give us equal control over its motion, we say that it has twice the mass of the latter car. The law of gravitation asserts that the gravitational pulls on two bodies are in exact proportion to their masses, so that if the earth's attraction on two bodies proves to be the same, their "masses" must be the same, whence it follows that the simplest way of measuring the mass of any body is by weighing it.

In the course of time, chemistry shewed that the Lucretian "atoms" had no right to their name (*ἀ-τέμνειν*, incapable of being cut). They proved not to be "uncuttable" at all, and so were henceforth called "molecules," the name "atom" being reserved for the smaller units into which the molecules could be broken up. There are many ways in which molecules may be broken up and their atoms re-arranged. Mere contiguity with other molecules may suffice, as for instance when iron rusts or acid is poured on to metal. Molecules may also be broken up by burning, exploding, heating, or by the incidence of light. For instance, if a bottle of hydrogen peroxide is stood in a light place, the mere passage of light through the liquid breaks up each molecule of hydrogen peroxide (H_2O_2) into a molecule of water (H_2O) and an atom of oxygen (O). When we take the cork out of our bottle we shall hear a "pop" caused by the escape of the oxygen gas, and find that some of the hydrogen peroxide has been changed into water. Molecules of silver bromide are also re-arranged by the incidence of light, this change forming the basis of photography.

Towards the end of the eighteenth century Lavoisier believed he had found that the total weight of matter remained unaltered throughout all the chemical changes at his command. In due course the law of "conservation of mass" became accepted as an integral part of science. We know now that it is not altogether exact; the weight of the oxygen which escapes from our bottle of peroxide, added to that of the fluid which remains, is slightly greater than the weight of the original peroxide, and a photographic plate gains in weight by being exposed to the light. We shall see shortly that the law is inexact because it neglects the weight of the light absorbed by the molecules of hydrogen peroxide or silver bromide.

The third principle, that of the conservation of energy, is the most recent of all. Energy can exist in a vast variety of forms, of which the simplest is pure energy of motion—the motion of a train along a level track, or of a billiard-ball over a table. Newton had shewn that this purely mechanical energy is "conserved." For instance, when two billiard-balls collide, the energy of each is changed, but the total energy of the two remains unaltered; one gives energy to the other, but no energy is lost or gained in the transaction. This, however, is only true if the balls are "perfectly elastic," an ideal condition in which the balls spring back from one another with the same speed with which they approached. Under actual conditions such as occur in nature, mechanical energy invariably appears to be lost; a bullet loses speed on passing through the air, and a train comes to rest in time if the engine is shut off. In all such cases heat and sound are produced. Now a long series of investigations has shewn that heat and sound are themselves forms of energy. In a classical series of experiments made in 1840–50, Joule measured the energy of heat, and tried to measure the energy of sound with the rudimentary apparatus of a violoncello string. Imperfect though his experiments were, they resulted in the recognition of "conservation of energy" as a principle which covered all known transformations of energy through its various modes of mechanical energy, heat, sound, and electrical energy. They shewed in brief that energy is transformed rather than lost, an apparent loss of energy of motion being compensated by the appearance of an exactly equal energy of heat and sound; the energy of motion of the rushing train is replaced by the equivalent energy of the noise of the shrieking brakes and of the heating of wheels, brake-blocks and rails.

Throughout the second half of the nineteenth century these three conservation laws stood unchallenged. The conservation of mass was supposed to be the same thing as the conservation of matter, because the mass of any body was regarded as the sum of the masses of its atoms; this of course explained simply—all too simply, as we now know—why total mass could not be altered by chemical action. But the newly discovered principle of conservation of energy stood apart from the two older laws, a thing by itself. The universe was still envisaged as a stage in which the players were atoms, each of which conserved its identity and mass through all time. To complete the picture, an entity known as energy was bandied about from one player to another, and this, like the actors themselves, was incapable of either creation or annihilation.

These three conservation laws ought of course to have been treated merely as working hypotheses, to be tested in every conceivable way and discarded as soon as they shewed signs of failing. Yet so securely did they seem to be established that they were treated as indisputable universal laws. Nineteenth-century physicists were accustomed to write of them as though they governed the whole of creation, and on this basis philosophers dogmatised as to the fundamental nature of the universe.

It was the calm before the hurricane. The first rumble of the approaching storm was a theoretical investigation by Sir J. J. Thomson, which shewed that the mass of an electrified body could be changed by setting it into motion; the faster such a body moved the greater its mass became, in opposition to Newton's concept of a fixed unalterable mass. For the moment, the principle of conservation of mass appeared to have abandoned science.

For a time this conclusion remained of merely academic interest; it could not be tested observationally because ordinary bodies could neither be charged with sufficient electricity, nor set into motion with sufficient speed, for the variations of mass predicted by theory to become appreciable in amount. Then, just as the nineteenth century was drawing to a close, Sir J. J. Thomson and his followers began to break up the atom, which now proved to be no more uncuttable, and so no more entitled to the name of "atom," than the molecule to which the name had previously been attached. They were only able to detach small fragments, and even now the complete break-up of the atom into its ultimate constituents has not been fully achieved. These fragments were found to be all precisely similar, and charged with negative electricity. They were accordingly named "electrons."

These electrons are far more intensely electrified than an ordinary body can ever be. A gramme of gold beaten, as thin as it will go, into a gold leaf a yard square, can with luck be made to hold a charge of about 60,000 electrostatic units of electricity, but a gramme of electrons carries a permanent charge which is about nine million million times greater. Because of this, and because electrons can be set into motion by electrical means with speeds of more than a hundred thousand miles a second, it is easy to verify that an electron's mass varies with its speed. Exact experiments have shewn that the variation is precisely that predicted by theory.

Thanks mainly to the researches of Rutherford, it has now been established that every atom is built up entirely of negatively charged electrons, and of positively charged particles called "protons"; matter proves to be nothing but a collection of particles charged with electricity. With one turn of the kaleidoscope all the sciences which deal with the properties and structure of matter have become ramifications of the single science of electricity. Before this, Faraday and Maxwell had shewn that all radiation was electrical in its nature, so that the whole of physical science is now comprised within the single science of electricity.

Since every body is a collection of electrically charged particles, the theoretical investigation already mentioned shews that the mass of every moving body must vary with its speed of motion. The mass of a moving body may be regarded as made up of two parts—a fixed part which the body retains even when at rest, known as its "rest-mass," and a variable part which depends on the speed of its motion. Both observation and theory have shewn that this second part is exactly proportional to the energy of motion of the body; the masses of two electrons, or any two other bodies similar to one another, differ to just the extent to which their energies differ.

In 1905 Einstein extended this into a tremendous generalisation. He shewed that not only energy of motion but energy of every conceivable kind must possess mass of its own; if it were not so, the theory of relativity could not be true. In this way every observational test of the theory of relativity was made a witness to the truth of the hypothesis that energy possesses mass. Einstein's investigation shewed that the mass of energy of any kind whatever depends solely on the amount of the energy, to which it is exactly proportional. It is also exceedingly small. The *Mauretania* fully loaded, weighs about 50,000 tons; when she is travelling at 25 knots, her motion only increases her weight by about a millionth part of an ounce. The energy that a man puts into a long life-time of heavy manual labour weighs only a 60,000th part of an ounce.

This discovery made it possible to reinstate the principle of conservation of mass. For mass is the aggregate of rest-mass and energy-mass, and as each of these is conserved separately (the former because matter is conserved, and the latter because energy is conserved), there must be a conservation of total mass. Nineteenth-century physics had regarded the conservation of mass as a consequence solely of the conservation of matter. Twentieth-century physics discovered that the conservation of energy was also involved; mass is now seen to be conserved only because matter and energy are conserved separately.

So long as atoms were regarded as permanent and indestructible—"the imperishable foundation-stones of the universe," to use Maxwell's phrase—it was natural to treat them as the fundamental constituents of the universe. The universe was, in brief, a universe of atoms, radiation being of quite secondary importance. Every now and then an atom was supposed to be set in vibration, as a bell is struck, and emitted radiation for a brief time, as a bell emits sound, until it lapsed back to its normal state of quiescence. But radiation was no more regarded as a primary constituent of matter than sound is of a carillon of bells. Incidentally this explains why it was found impossible to imagine how the sun could continue to radiate for thousands of millions of years or more. Sunlight was believed to be produced by the agitation of atoms, but no one could imagine what maintained the agitation.

The scene began to change as soon as it was recognised that the atom was built up of electrified particles. For no matter how far we retreat from an electrified particle, we cannot get outside the range of its attractions and repulsions. This shews that an electron must, in a certain sense at least, occupy the whole of space. Faraday and Maxwell made the matter more explicit than this; they pictured an electrified particle as an octopus-like structure, a small concrete body which threw out a sort of feelers or tentacles, called "lines of force," throughout the whole of space. When two electrified particles attracted or repelled one another, it was because their tentacles had somehow taken hold of one another, and pushed or pulled. These tentacles were supposed to be formed out of electric and magnetic forces, of which radiation is also formed. When an atom emitted radiation it merely discharged some of its tentacles into space, much as a porcupine is said to throw out its quills. This concept placed radiation and matter in more intimate relations than ever before.

Since all types of radiation are forms of energy, they must, in accordance with Einstein's principle, carry mass associated with them. When an atom emits radiation, its mass diminishes by the mass of the emitted radiation, just as, if a porcupine were to throw out its quills, its weight would diminish by the weight of the quills. Thus when a piece of coal is burnt, its weight is not altogether reproduced in the ashes and the smoke; we must add to these the weight of the light and heat emitted in the process of combustion. Only then will the total be exactly the weight of the original piece of coal.

As far back as 1873, Maxwell had shewn that radiation would exert a pressure on any surface on which it fell. We now regard this as a necessary consequence of the fact that radiation carries mass about with it; a beam of light consists of mass moving with the speed of light—186,000 miles a second. Subsequently Lebedew observed this pressure, and Nichols found its amount to be that calculated by Maxwell. A target could be seen to flinch under the impact of the radiation from a bright light, just as though a bullet had been fired into it. But the impact of such light as we experience on earth is extremely slight; to see the full implications of the phenomenon we must leave the earth and the physics which has been developed in terrestrial laboratories, in favour of the sky and the wider physics which we see in operation in the colossal crucibles of the stars. Heat an ordinary six-inch cannon-ball up to 50 million degrees, which is the kind of temperature we expect to find at the centre of the sun or of an average star, and the radiation it emits would suffice to mow down—by its mere impact, like the jet of water from a fire-hose—anyone who approached within 50 miles of it. Indeed inside the stars this pressure of radiation is so large that it supports an appreciable fraction of the weight of the stars.

Calculation shews that about a ten-thousandth of an ounce of sunlight falls every minute on every square mile of land directly under the sun; it falls with the speed of light, and in being brought to rest it exerts a pressure of about 0·000,000,000,04 atmosphere on the land. The figures look absurdly small—the weight of sunshine which falls in a century is less than the weight of rain which falls in a fiftieth of a second of a heavy shower. Yet the amount is small only because a field a mile square is such a minute object in astronomical space. The total emission of radiation by the sun is almost exactly 250 million tons a minute, which is something like 10,000 times the average rate at which water flows under London Bridge. And, incidentally, if our factor of 10,000 is wrong, it is not because we do not know the exact weight of solar radiation, but because we do not know the average flow of the Thames with very great precision. Astronomical physics is a far more exact science than terrestrial hydraulics.

A certain weight of radiation falls on to the sun from other stars, but this is quite inappreciable in comparison with the weight of the radiation which streams out, so that the sun can only maintain its weight if actual matter is streaming into it at the rate of close upon 250 million tons a minute.

As the sun journeys through space it must continually sweep up stray matter in the form of odd atoms and molecules, of dust particles and of meteors. These last are small solid objects which exist in enormous numbers in the solar system, revolving around the sun in orbits like those of the planets. Occasionally they dash into the earth's atmosphere, when the air-resistance of their earthward fall raises them to incandescence, and they appear as shooting-stars. Generally these dissolve into vapour before reaching the earth's surface; only occasionally is one massive enough to survive the disintegrating effect of this air-resistance, and it then strikes the earth in the form of a stone, known as a meteorite. These are sometimes of enormous size. The fall of a meteorite in Siberia in 1908 set up blasts of air which devastated the forests over an enormous area, while the shock of its impact on the solid earth caused waves which were recorded thousands of miles away. And a vast crater-shaped depression in Arizona, three miles in circumference, is believed to have been caused by the fall of a still larger meteorite in prehistoric times. Yet such giants are rare, and the average meteor is a puny affair, generally no larger than a cherry or a pea.

Shapley has estimated that many thousands of millions of shooting-stars enter the earth's atmosphere every day; each of these is turned into dust and vapour, and the earth's weight is correspondingly increased. An incomparably greater number must fall into the sun, measured by millions of millions per second, and these probably provide by far the largest contribution to the sun's bag of stray matter. Yet Shapley estimates that the total weight of meteoric matter falling into the sun can hardly exceed 2000 tons a second, which is less than a 2000th part of the weight it loses by radiation. Thus it seems fairly certain that on the balance the sun must be losing weight at a rate of very near 250 million tons a minute; it is a wasting structure, gradually disappearing before our eyes; it is melting away like an ice-berg in the Gulf Stream. And the same must be true of other stars.

This conclusion accords well with the general broad facts of astronomy. Although there is no absolute proof, a large accumulation of evidence goes to shew that young stars are heavier than old stars. They are not heavier merely by a few million tons, but several times heavier—often 10, 50 or even 100 times heavier. By far the simplest explanation is that the stars lose the greater part of their weight in the course of their lives. Now a simple calculation shews that the sun, losing weight at a rate of about 250 million tons a minute, would require millions of millions of years to lose the greater part, or even a considerable part, of its weight. And, as other stars tell much the same story, we are led to assign lives of millions of millions of years to the stars in general.

We have other means of estimating the length of stellar lives. In particular, the motion of the stars in space proclaims their extreme antiquity, and again assigns to them lives of millions of millions of years. We have seen how far removed from one another in space the stars are—so far that it is very rare for two stars to approach each other at all closely. Yet if the stars have lived these tremendously long lives of millions of millions of years, each star ought to have experienced a number of fairly close approaches. The gravitational pulls which the stars would exert on one another on these occasions would not generally be intense enough to tear out planets, but would suffice to deflect the stars from their courses and change the speeds of their motions. In the case of binary systems, which consist of two separate masses moving through space in double harness like a single star, the gravitational pull of a near star would re-arrange the orbits of the two constituents of the binary star.

Now all these effects can be calculated in detail, so that we know exactly what to expect if the stars have really lived the terrifically long lives of millions of millions of years we are provisionally allotting to them. And everything we look for we find. All the anticipated effects are there, and, so far as we can tell, their magnitudes indicate that the stars have lived for millions of millions of years.

Against all this, there is evidence of another kind, which seems to point to a very different conclusion, and so must be discussed in some detail even though it is highly technical, and takes us into the most difficult parts of the difficult theory of relativity.

As we shall see in the next chapter, this theory tells us that space itself is curved, much in the same way in which the surface of the earth is curved. The curvature of space is responsible for the curving of rays of light which is observed at a solar eclipse, and for the curvature in the paths of planets and comets, which we used to attribute to a "force" of gravitation. On this theory, the presence of matter does not produce "force," which is an illusion, but a curving of space. To confront our difficulties singly, let us for the moment suppose that the presence of matter is the only cause of the bending of space. Then an empty universe, totally devoid of matter, would have its space entirely uncurved, because there would be no matter to curve it, and so would be of infinite size. As the universe is not empty, its size will be determined by the amount of matter it contains. The more matter there is in the universe, the more curved space will be, the more rapidly it will bend back on itself, and as a consequence the smaller the universe will be—just as a circle which curves rapidly is smaller than one which curves more gradually.

The well-known experiment of electrifying a soap-bubble may make the concept clearer. A soap-bubble, blown in the ordinary way, is allowed to rest on the plate of an electrical machine. As the machine is worked, and the bubble becomes more and more highly charged with electricity, its size increases steadily until finally it bursts. Here (apart from its final bursting) the soap-bubble is analogous to the universe; its size depends on the amount of electricity it carries, just as the size of the universe depends on the amount of matter it contains. And yet there are two essential differences. The first is that a soap-bubble has a certain curvature inherent in its structure, so that it is of definite and finite size, even when uncharged; the universe, on the other hand, becomes infinite in size when it is empty of matter. The second is that increasing the charge of electricity *increases* the size of the soap-bubble, but increasing the amount of matter *decreases* the size of the universe—the more matter there is, the less space there is to hold it.

Einstein tried to obviate this last objection, as well as others, by making the universe more like the soap-bubble. He imagined it to have an inherent curvature, besides that produced by matter, of such a kind that its size would *increase* if the amount of matter increased.

Even so, there is still one outstanding difference. The gravitating masses in space all attract one another, but the electric charges on the soap-bubble repel one another, because they are all of similar electricity, whether positive or negative. As a consequence of this, the electrified soap-bubble is a thoroughly stable structure. Add a little more charge and it calmly adjusts itself to a new, slightly expanded, position of equilibrium. Shake it, and, after trembling for a bit, it settles down to rest again. But, just because of the difference between attraction and repulsion, a soap-bubble charged with attracting matter would be unstable. The mathematician will see why this must be so. And although it is a long step from a two-dimensional soap-bubble of liquid film to a universe, a recent investigation by a Belgian mathematician, the Abbé Lemaître, has shewn that the analogy holds, and that the kind of universe we have just been discusssing would be an unstable structure; it could not stay at rest for long, but would start at once to expand to infinite size or contract to a point. Hence the actual space of an aged universe ought to be either expanding or contracting, and the various objects in it all rushing away from one another, or all rushing towards one another, at a great rate.

Lemaître's conclusions are based upon Einstein's concept of a universe whose size, when at rest, depends on the amount of matter it contains. Previously to this, however, a very different concept of the universe had been put forward by Professor de Sitter of Leiden. Like Einstein, he supposed the universe to possess a certain amount of curvature, impressed upon it by the inherent properties of space and time. The presence of matter added an additional curvature, but, as matter is so sparsely distributed in the actual universe, this was insignificant in comparison with the curvature resulting from the nature of space and time. When de Sitter studied the properties of his universe mathematically, he too found a tendency for its space to expand or contract, and for all the objects in it either to drift apart or to rush towards one another.

At first de Sitter's concept of the universe appeared to be entirely antagonistic to Einstein's earlier concept, and mathematicians were content to wait for something to decide between them. But Lemaître's work now shews that the two concepts are not so much competitive as complementary. As Einstein's unstable universe expands, the matter in it becomes more and more sparse until it ends up as an empty universe of the kind pictured by de Sitter. The universes of Einstein and de Sitter may rightly be imagined as placed at the two ends of a chain, but we shall go wrong if we imagine them engaged in a tug-of-war. They merely mark the limits of possible universes, and a universe which starts at or near the Einstein end of the chain must gradually slip along the chain to the de Sitter end. If our universe is built on these lines at all, the question before us is not at which end of the chain it is, but how far along the chain it has travelled.

The two ideal universes at the two ends of the chain are similar in that the objects in them must be either all rushing away from one another or else all rushing towards one another. This is not only true at the two extreme ends of the chain, but all along the chain. If the universe is built in accordance with the theory of relativity, as it almost certainly is, then the objects in it must be running all away from one another or all towards one another.

These conclusions are of great interest, because it has for some years been remarked that the remote spiral nebulae are, to all appearances, rushing away from the earth, and so presumably also from one another, at terrific speeds, which become greater and greater the farther we recede into space. The last nebula investigated at Mount Wilson—one of the most distant which can be observed in the great 100-inch telescope—was found to be receding at the terrific speed of 15,000 miles a second. Dr Hubble and Dr Humason, who have made a special study of the question at Mount Wilson, find that the speeds at which the individual nebulae are receding from us are, roughly speaking, proportional to their distances from us, as they ought to be, if the cosmology of the theory of relativity is correct. A nebula whose light takes ten million years to reach us, has a speed of about 900 miles a second, and the speeds of other nebulae are, approximately at least, proportional to their distances. For instance, the light from the nebulae shewn in Plate I takes 50 million years to reach us, and the nebulae shew speeds of recession of about 4500 miles a second.

The actual figures are important, because if we trace the implied nebular motions backwards, we find that all the nebulae must have been congregated in the neighbourhood of the sun only a few thousands of millions of years ago. All this goes to suggest that we are living in an expanding universe, which started to expand only a few thousands of millions of years ago.

If this were the whole story, it would be very difficult to assign ages of millions of millions of years to the stars; this would imply that they had been packed close together, or had been converging into a small region of space, for millions of millions of years, and only just recently, during the last thousandth part or so of their existence, had begun to scatter. If the supposed motions of recession ultimately prove to be real, it will hardly be possible to attribute an age of more than a few thousands of millions of years to the universe.

But there is room for a good deal of doubt as to whether these huge speeds are real or not. They have not been obtained by any direct process of measurement, but are deduced by an application of what is known as Doppler's principle. It is a matter of common observation that the noise emitted by a motor-car horn sounds deeper in pitch when it is receding from us than when it is coming towards us. On the same principle the light emitted by a receding body appears redder in colour than that emitted by a body approaching us, colour in light corresponding to pitch in sound. By accurately measuring the colour of well-defined spectral lines, the astronomer is able to discover whether the body emitting them is approaching us or receding from us, and can estimate the speed of the motion. And the only reason for thinking that the distant nebulae are receding from us is that the light we receive from them appears redder than it ought normally to be.

Yet other things than speed are capable of reddening light; for instance, sunlight is reddened by the mere weight of the sun, it is reddened still more by the pressure of the sun's atmosphere; it is further reddened, although in a different way, in its passage through the earth's atmosphere, as we see at sunrise or sunset. The light emitted by certain stars of a different kind is reddened in a mysterious way we do not yet understand. Furthermore, on de Sitter's theory of the universe, distance alone produces a reddening of light, so that even if the distant nebulae were standing still in space, their light would appear unduly red, and we should be tempted to infer that they were receding from us. None of these causes seems capable of explaining the observed reddening of nebular light, but quite recently Dr Zwicky of the California Institute has suggested that still another cause of reddening may be found in the gravitational pull of stars and nebulae on light passing near them —the same pull as causes the observed bending of starlight at an eclipse of the sun. Compton's experiments (p. 34) shew that radiation is both deflected and reddened when it encounters electrons in space. When radiation interacts gravitationally with stars or other matter in space, it is known to be deflected, and Zwicky's suggestion is that it is reddened as well.

To test this suggestion, ten Bruggencate has examined the light from a number of globular clusters, all at about equal distances from us, but so selected that the amount of intervening gravitational matter varied greatly. The light from these shewed a reddening, and if this were caused by the expansion of space, it ought to have been the same for all the clusters. Actually it proved to be far from uniform; it was much more nearly proportional to the amount of intervening matter, exactly as required by Zwicky's theory, and its actual amount agreed well enough with that predicted by the theoretical formula. As we can hardly imagine that the globular clusters, which belong to our own galactic system of stars, can be systematically running away from us, the case for supposing that the spiral nebulae are running away becomes very much weaker, Zwicky's theory providing a possible explanation of the observed reddening of the light.

Other lines of evidence also suggest that the suspected recessions of the nebulae may be spurious. For instance the light from the nearest nebulae is not redder but bluer than normal, and as light can only be made bluer by an actual physical approach, this can only mean that the nearest nebulae are actually coming towards us. Moreover, the apparent speeds of the nebulae are by no means strictly proportional to their distances; for instance, nebulae believed to be at the same distance of seven million light years shew deviations averaging 240 miles a second out of total speeds of 640 miles a second.

Nevertheless, if the universe is built in the way we have described, the nebulae as a whole must undoubtedly be running away from us; theoretical considerations demand this and cannot be satisfied with anything less, but they do not tell us the speeds of the nebular motions. The work of Zwicky and ten Bruggencate in no way throws doubt on there being a real motion of recession; what it lays open to doubt is whether this motion is the same as astronomers have deduced from the reddening of the spectral lines. Possibly most of this reddening may be attributed to the effect suggested by Zwicky, or to some similar cause, while only a small residual represents a real motion of recession. It is impossible to determine the speed of this motion because the smaller effect is entirely masked by the greater.

The question is still an open one, but if once it is accepted that the greater part of the apparent velocities of recession may be treated as spurious, the argument in favour of short lives for the stars disappears, and we become free to assign to them the long lives of millions of millions of years which the general evidence of astronomy seems to demand.

As we have already seen, this general evidence suggests that the sun has been pouring away mass in the form of radiation at a rate of 250 million tons a minute for a period of some millions of millions of years. Detailed calculation shews that the new-born sun must have had many times the mass of the present sun, in conformity with the general fact of observation that young stars are many times more massive than old stars. In what form could it store all the mass which has since disappeared in the form of radiation?

The rest-mass of an electron or other charged particle is generally enormously greater than its energy-mass, the latter assuming its greatest importance at high temperatures. Now the temperature at the centre of the sun is about 50,000,000 degrees, and even here the rest-mass accounts for all but about one part in 200,000 of the total mass. It is improbable that the new-born sun can have been much hotter than this, so that it seems likely that the greater part of the mass of the primaeval sun also must have resided in its rest-mass. If so there is only one conclusion possible: the primaeval sun must have contained many more electrons and protons, and therefore many more atoms, than now. These atoms can only have disappeared in one way: they must have been annihilated, and their mass must be represented by the mass of the radiation which the sun has emitted in its long life of millions of millions of years.

This argument may be thought somewhat precarious, because it deals with concepts so far out of the range of laboratory physics. Fortunately laboratory physics has quite recently obtained evidence, which, although far from being absolutely conclusive, provides valuable confirmation that this annihilation of matter is actually taking place on a vast scale out in the depths of space.

We could hardly expect to obtain direct evidence of the annihilation of matter going on in stellar interiors, because the radiation produced in the process could only travel a very short distance before being absorbed by the substance of the star. This would be heated up, and the corresponding energy would ultimately be emitted by the star in the form of quite ordinary light and heat.

A mathematical analysis of the facts of astronomy suggests that the process of atomic annihilation would probably be spontaneous in the same way in which radioactive disintegration is spontaneous. If so, it would not be limited to the hot interiors of stars, but ought to be in progress wherever astronomical matter exists in sufficient abundance.

In its simplest form the process would consist of the

simultaneous annihilation of a single electron and a single proton. We can picture it vividly if we think of these two charged particles rushing together under their mutual attraction with ever-increasing speed, until finally they coalesce; their electric charges then neutralise one another, and their combined energy is set free in a single flash of radiation—a "photon" of the kind discussed on p. 32.

We have already seen (p. 48) how mass is "conserved" when an atom emits radiation. The atom parts with a certain amount of its mass, but this is not destroyed; it is carried away by the photon, and figures as the mass of the photon. If a proton and electron annihilate one another, the resulting photon must have a mass equal to the combined masses of the proton and electron which have disappeared. Now the combined mass of a proton and electron is known with great accuracy, for it is exactly equal to the mass of the hydrogen atom. Thus if the annihilation of matter really occurs, photons of mass exactly equal to that of the hydrogen atom ought to be traversing space in great numbers, and some of these ought to fall on the earth.

There may be even more massive photons than this, for we can imagine any kind of atom being suddenly annihilated, and setting loose its whole energy as a photon, whose mass would then be equal to that of the whole atom. One possibility is of special interest. Although we believe that all matter is in the last resort built up of protons and electrons, there is a peculiarly compact structure of four protons and two electrons which may almost be considered as a new and independent unit. It is conspicuous in the radiation emitted by radio-active substances, and is commonly known as an α-particle. The helium atom, which is the next simplest atom after hydrogen, consists of an α-particle with two electrons revolving in orbital motion about it. As an α-particle has the same electric charge as two protons, it might undergo annihilation by coalescing with two electrons, in which event the resulting photon would have the same mass as a helium atom.

Photons of either of these two kinds would have an incomparably greater mass than the photons of any ordinary kind of radiation, and so ought to be immediately recognisable. Photons may be regarded as bullets, all travelling with a uniform speed—the speed of light. If a number of bullets are discharged from a gun with equal speeds, the more massive projectiles will have the greater capacity for doing damage, and so will have the greater penetrating power. It is the same with a mixed crowd of photons; the more massive photons have the greater penetrating power. There is a mathematical formula which enables us to deduce the penetrating power of a photon from its mass, and it shews that photons having the mass of atoms of either hydrogen or helium ought to have terrific powers of penetration.

We have already spoken of the highly penetrating radiation, commonly called "cosmic radiation," which falls on the earth from outer space, and is able to penetrate several yards of lead. For a long time it was not altogether clear whether this was a true radiation, or consisted of streams of electrons. The former alternative always seemed by far the more probable, because electrons would have to move with almost unthinkably high energy to force their way through many yards of lead before being brought to rest.

The matter now appears to be settled. A shower of

electrons, falling on to the earth from outer space, would become entangled in the earth's magnetic field, and this would influence its motion. If the electrons were moving fast enough to have the observed penetrating power of cosmic radiation, calculation shews that almost the whole stream would be deflected from its course, and strike the earth near to one or other of its magnetic poles. No such preference is shewn by the cosmic rays; different observers, working at different parts of the earth's surface, find that the radiation has the same intensity everywhere. For instance, the British Australian and New Zealand Antarctic Expedition found the same intensity within 250 miles of the south magnetic pole as other observers had found in regions remote from the poles. This makes it reasonably certain that the "cosmic radiation" is true radiation, and not merely a shower of electrons. This being so, we can deduce the mass of the photons of the radiation from their observed penetrating power by the use of the formula already mentioned.

The penetrating power of this radiation has been studied with extreme care and skill by Professor Millikan and his colleagues at Pasadena, by Professor Regener of Stuttgart, and by many others. They all find that the radiation is a mixture of a number of constituents of very different penetrating powers, or, what is the same thing, a mixture of photons of different masses. Now it seems highly significant that the two ingredients of highest penetrating power consist of photons whose masses are, as nearly as we can tell, equal to the masses of the helium atom and the hydrogen atom respectively; in other words they are just the type of photons we should expect to find if, somewhere out in the far depths of space, protons and α-particles were being annihilated, the former in conjunction with the single electrons needed to neutralise their charges, and the latter in conjunction with the pairs of electrons needed for the same purpose.

It must be explained that the masses of the photons cannot be measured with absolute precision, so that it cannot be claimed with certainty that they are absolutely and precisely those to be expected from the annihilation in question. Yet the agreement is about as good as observation permits; in each case there is agreement to within about 5 per cent., and the penetrating power of the radiation can hardly be measured more closely than this. Such an agreement is too good to be dismissed as a mere coincidence, so that it seems highly probable that this radiation has its origin in the actual annihilation of protons and electrons.

Nevertheless, the matter is not yet beyond controversy, and the view I have just stated is not universally accepted by physicists. Professor Millikan, in particular, has suggested that cosmic radiation may originate in the process of building up heavy atoms out of simpler light atoms, and so interprets it as evidence that "the creator is still on the job." To take the simplest illustration, a helium atom contains exactly the same ingredients as four hydrogen atoms—namely, four electrons and four protons—but its mass is only equal to that of 3·97 hydrogen atoms. Thus if four hydrogen atoms could somehow be hammered together to form a helium atom, the superfluous mass, that of 0·03 hydrogen atoms, would take the form of radiation, and a photon with 3 per cent. of the mass of the hydrogen atom might be discharged. We cannot say it would be discharged, because if ever four hydrogen atoms fall together to form a helium atom, it seems likely that the process would occur in several stages, and so would result in the emission of a number of small photons rather than of one big one. Yet even if the whole of the liberated energy were to form one big photon, this would have less penetrating power than the actual cosmic radiation. If, however, 129 atoms of hydrogen were to fall together and form a single atom of xenon by one huge cataclysmic disturbance, the single photon emitted in the process would have about the same mass as the hydrogen atom, and so would have something like the same penetrating power as the second most penetrating constituent of actual cosmic radiation. On this view of the origin of the radiation, the less penetrating constituents can be very readily and naturally explained as originating out of the synthesis of atoms less complex than xenon. On the other hand, the most penetrating constituent of all seems to present a quite insuperable difficulty. If its photons originate out of the hammering together of hydrogen atoms to form a single huge atom, this atom must needs have an atomic weight in the neighbourhood of 500, which seems beyond the bounds of probability. It seems almost equally improbable that the second most penetrating constituent should be produced by the synthesis of atoms of xenon or other element of similar atomic weight, since all such atoms are of extreme rarity. Whatever the origin of the less penetrating constituents, the two most penetrating constituents can hardly, I think, be attributed with much plausibility to any other source than annihilation of matter.

The amount of this radiation which falls on the earth is tremendous. Millikan and Cameron have estimated it at about a tenth of the total radiation received from all the stars in the sky, the sun of course excepted. Out in the depths of space, beyond the Milky Way, the highly penetrating radiation must still be about as plentiful as it is at the earth's surface, but starlight is far less plentiful, so that, on taking an average through space as a whole, this highly penetrating radiation is probably the commonest kind of radiation.

Its vast amount is explained in part by its high penetrating power, which almost endows it with immortality. An average beam of the radiation travelling through space for millions of millions of years will not encounter matter to absorb it to any appreciable extent. Thus we must think of space as being drenched with almost all the cosmic radiation which has ever been generated since the world began. Its rays come to us as messengers not only from the farthest depths of space, but also from the farthest depths of time. And, if we read it aright, their message seems to be that somewhere, sometime, in the history of the universe, matter has been annihilated, and this not in tiny, but in stupendous amounts.

If we accept the astronomical evidence of the ages of the stars and the physical evidence of the highly penetrating radiation as jointly establishing that matter can really be annihilated, or rather transformed into radiation, then this transformation becomes one of the fundamental processes of the universe. The conservation of matter disappears entirely from science, while the conservation of mass and of energy become identical. Thus the three major conservation laws, those of the conservation of matter, mass and energy, reduce to one. One simple fundamental entity which may take many forms, matter and radiation in particular, is conserved through all changes; the sum total of this entity forms the whole activity of the universe, which does not change its total quantity. But it continually changes its quality, and this change of quality appears to be the main operation going on in the universe which forms our material home. The whole of the available evidence seems to me to indicate that the change is, with possible insignificant exceptions, for ever in the same direction—for ever solid matter melts into insubstantial radiation: for ever the tangible changes into the intangible.

These concepts have been discussed at some length because they obviously have a very special bearing on the fundamental structure of the universe. In the last chapter we saw how the wave-mechanics reduced the whole universe to systems of waves. Electrons and protons consisted of waves of one kind; radiation of waves of a different kind. The discussion of the present chapter has suggested that matter and radiation may not constitute two distinct and non-interchangeable forms of waves. The two may be interchangeable, one passing into the other as the chrysalis passes into the butterfly—to which, as we shall see below (p. 133), some scientists might think it necessary to add "and as we can imagine the butterfly to pass back into the chrysalis."

This does not of course mean that matter and radiation are the same thing. The transformation of matter into radiation still means something, although the concept now looks incomparably less revolutionary than it looked when first I advanced it twenty-six years ago. Even if we knew all the facts with certainty, which we do not, it would be difficult to express the situation accurately in non-technical language, but possibly we may come fairly near to the truth if we think of matter and radiation as two kinds of waves—a kind which goes round and round in circles, and a kind which travels in straight lines. The latter waves of course travel with the velocity of light, but those which constitute matter travel more slowly. It has even been suggested, by Mosharrafa and others, that this may express the whole difference between matter and radiation, matter being nothing but a sort of congealed radiation travelling at less than its normal speed. We have already seen (p. 38) how the wave-length of a moving particle depends on its speed. The dependence is such that a particle travelling with the speed of light would have precisely the same wave-length as a photon of equal mass. This remarkable fact, as well as others, goes a long way towards suggesting that radiation may ultimately prove to be merely matter moving with the speed of light, and matter to be radiation moving with a speed less than that of light. But science is a long way from this as yet.

To sum up the main results of this and the preceding chapter, the tendency of modern physics is to resolve the whole material universe into waves, and nothing but waves. These waves are of two kinds: bottled-up waves, which we call matter, and unbottled waves, which we call radiation or light. The process of annihilation of matter is merely that of unbottling imprisoned wave-energy and setting it free to travel through space. These concepts reduce the whole universe to a world of radiation, potential or existent, and it no longer seems surprising that the fundamental particles of which matter is built should exhibit many of the properties of waves.

Chapter IV

RELATIVITY AND THE ETHER

We have seen how modern physics reduces the universe to systems of waves. If we find it hard to imagine waves unless they travel through something concrete, let us say waves in an ether or ethers. I believe it was the late Lord Salisbury who defined the ether as the nominative of the verb "to undulate." If this definition will serve for the moment, we can have our ether without committing ourselves very far as to its nature. And this makes it possible to sum up the tendency of modern physics very concisely: modern physics is pushing the whole universe into one or more ethers. It will be well, then, to scrutinise the physical properties of these ethers with some care, since in them the true nature of the universe must be hidden.

It may be well to state our conclusion in advance. It is, in brief, that the ethers and their undulations, the waves which form the universe, are in all probability fictitious. This is not to say that they have no existence at all: they exist in our minds, or we should not be discussing them; and something must exist outside our minds to put this or any other concept into our minds. To this something we may temporarily assign the name "reality," and it is this reality which it is the object of science to study. But we shall find that this reality is something very different from what the scientist of fifty years ago meant by ether, undulations and waves, so much so that, judged by his standards and speaking his language for a moment, the ethers and their waves are not realities at all. And yet they are the most real things of which we have any knowledge or experience, and so are as real as anything possibly can be for us.

The concept of an ether entered science some two centuries ago or more. When the known properties of gross matter failed to explain a phenomenon, scientists met the difficulty by creating a hypothetical all-pervading ether, to which they attributed exactly the properties necessary to provide an explanation. There was of course a special temptation to resort to this procedure in problems which appeared to call for "action-at-a-distance." It is, on the face of it, such good sound sense to assert that matter can only act where it is, and cannot possibly act where it is not, that he who argues to the contrary can hardly hope to carry the majority of his fellows with him. Descartes had gone so far as to argue that the bare existence of bodies separated by distance was a sufficient proof of the existence of a medium between them.

Thus when no gross material was present to transmit a mechanical action, such as that exerted by a magnet on a steel bar, or by the earth on a falling apple, the temptation to invoke an all-pervading ether became well-nigh irresistible, and what may be called the ether-habit invaded science. So that, as Maxwell expressed it: "Ethers were invented for the planets to swim in, to constitute electric atmospheres and magnetic effluvia, to convey sensations from one part of our body to another, till all space was filled several times over with ether." In the end there were almost as many ethers as unsolved problems in physics.

Fifty years ago only one of these ethers survived in serious scientific thought—the luminiferous ether, which was supposed to transmit radiation. The properties it needed to fulfil this function had been defined with ever-increasing precision by Huyghens, Thomas Young, Faraday and Maxwell. It was thought of as a jelly-like sea through which waves could travel, just as vibrations or undulations travel through a jelly. These waves were radiation which, as we now know, can take any one of the many forms of light, heat, infra-red or ultra-violet radiation, electromagnetic waves, X-rays, γ-rays, and cosmic radiation.

The astronomical phenomenon of the "aberration of light," as well as a number of others, shew that, if such an ether exists, the earth and all other moving bodies must pass through it without disturbing it. Or, if we take our position on the earth and study the phenomena from that standpoint, the ether must pass through the interstices of the earth and other solid bodies without hindrance—"like the wind through a grove of trees," to borrow the famous but inadequate simile of Thomas Young. It is inadequate because wind does in actual fact affect trees; the motions of their leaves, twigs and branches give some indication of its strength. But it can be shewn that motion through the ether cannot in the least degree disturb solid bodies which are at rest on the earth, or affect their motions if they are moving; we need not add ether-resistance to air-resistance in discussing what prevents our motor-car making better speed.

Thus, if an ether exists, it is all the same whether the ether-wind is blowing past us at one mile an hour or a thousand miles an hour. This is in accordance with a dynamical principle which Newton had enunciated in his *Principia*:

COROLLARY V: **The motions of bodies included in a given space are the same among themselves, whether that space is at rest, or moves uniformly forwards in a right line without any circular motion.**

Newton continues:

A clear proof of which we have from the experiment of a ship, where all motions happen after the same manner whether the ship is at rest, or is carried uniformly forward in a right line.

This general principle shews that no experiment performed on board ship and confined to the ship alone can ever reveal the ship's velocity through a still sea. Indeed it is a matter of common observation that in calm weather we cannot even tell in which direction a ship is moving without looking at the sea.

If the ether-wind had affected terrestrial bodies, the disturbance it created would have given an indication of the speed with which it was blowing, just as the motions of the twigs of trees give an indication of ordinary wind-velocity. As things are, it is necessary to resort to other methods.

Although an ocean traveller cannot determine the speed of his ship by any observation which is confined to the ship, he can easily do so if he is free to observe the sea as well. If he drops a line and sounding-lead into the sea, a circular ripple will spread out; but every sailor knows that the point at which the line enters the water will not remain at the centre of this circle. The centre of the circle stays fixed in the water, but the point of entry of the line is dragged forward by the motion of the ship, so that the rate at which the point of entry advances from the centre of the circle will disclose the speed of the ship through the sea.

If the earth is ploughing its way through a sea of ether, an experiment conceived on similar lines ought to reveal the speed of its progress. The famous Michelson-Morley experiment was designed to precisely this end. Our earth was the ship, and the physical laboratory of the University of Cleveland (Ohio) was the point of entry of the lead into the sea. The dropping of the lead was represented by the emission of a light-signal, and it was supposed that the light-waves which constituted this signal would make ripples on the sea of ether.

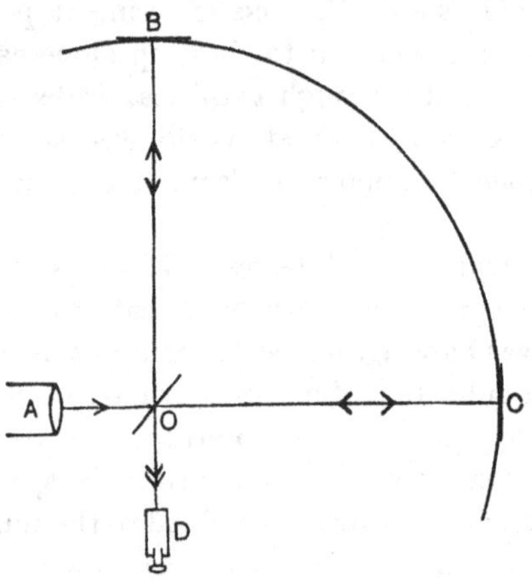

Fig. 1. Diagram to illustrate the Michelson-Morley experiment

Light from a source *A* is projected on to a half-silvered mirror *O*, so that half is reflected along *OB* and the rest continues along *OC*, of length equal to *OB*, actually about 12 yards. Mirrors at *B* and *C* reflect the light back to *O*, and half of each beam then passes into a small telescope *D*. The amount by which one lags behind the other is compared with the lag when the whole apparatus has been turned through 90°. This procedure eliminates any error caused by *OB* and *OC* being slightly different in length.

The progress of the ripples could not be followed directly, but sufficient information could be obtained by arranging for mirrors to reflect the signal back to the starting-point. This made it possible to determine in effect the time which the light took to perform the double journey to and fro. If the earth were standing still in the ether, the time of a double journey of given length would of course always be the same, regardless of its direction in space. But if the earth were moving through a sea of ether in an easterly direction, it is easy to see that a double journey, first from east to west and then from west to east, ought to take slightly more time than one of equal length in north-south and south-north directions. No more recondite principle is involved than in the common experience that it takes longer to row a boat 100 yards up-stream and 100 yards down-stream than to row 200 yards across the stream; in the former case we go slowly up-stream, and come quickly down-stream, but the gain of time in rowing down with the current is not sufficient to make good the time previously lost in rowing up against the current. If two oarsmen of equal speed set out simultaneously to row the two courses, the cross-stream rower will arrive first, and the difference between their times of arrival will disclose the speed of the current. It was anticipated that, in precisely the same way, the difference in the times taken by the two beams of light in the Michelson-Morley experiment would disclose the speed of the earth's motion through the ether.

The experiment was performed many times, but no time-difference at all could be detected. Thus, on the hypothesis that our earth was surrounded by a sea of ether, the experiments seemed to shew that its speed of motion through this sea of ether was zero. To all appearances, the earth stood permanently at rest in the ether, while the sun and the whole of creation circled round it; the experiments seemed to bring back the geocentric universe of pre-Copernican days. Yet it was impossible that this should be their true interpretation, for the earth was known to be moving round the sun at a speed of nearly twenty miles a second, and the experiments were sensitive enough to detect a speed of one-hundredth part of this.

Fitzgerald in 1893 and Lorentz independently in 1895 suggested an alternative interpretation. The experimenters had in effect tried to make two rays of light travel simultaneously to and fro over two courses of equal length. Without losing anything of the essence of the experiment, we may imagine that the lengths of the two courses had been measured or compared by ordinary measuring rods—foot-rules, if we like. How was it known, Fitzgerald and Lorentz asked, that these rods, or the course laid out by them, retained their exact length while they were moving forward through a sea of ether? When a ship moves through the ocean, the pressure of the sea on its bows causes it to contract its length; it is, so to speak, squeezed up a little bit—a minute fraction of an inch—between the sea trying to hold its bows back and its screw trying to push its stern forward. In the same way a motor-car moving through the air contracts as it is squeezed between the backward pressure of the wind on its windscreen, and the forward drive of its rear wheels. If the apparatus used by Michelson and Morley contracted in the same way, the up-and-down stream course would always be shorter than the cross-stream course. This reduction of length would do something to compensate for the other disadvantages of the up-and-down stream course. A contraction of exactly the right amount would compensate for them completely, so that this and the cross-stream course would require precisely equal times. In this way, Fitzgerald and Lorentz suggested, it might be possible to account for the *nul* result of the experiment.

The idea was not wholly fanciful or hypothetical, for

Lorentz shewed very shortly afterwards that the electrodynamical theory then current demanded that just such a contraction should actually occur. Although the contraction was not altogether analogous to those of ships or motor-cars, these give a good enough idea of the mechanism involved. Actually Lorentz shewed that if matter were a purely electrical structure, consisting solely of electrically charged particles, motion through the ether would cause the particles to readjust their positions, and they would not come to relative rest again until the body had contracted by a certain calculable amount. And this amount proved to be precisely that needed to account for the *nul* result of the Michelson-Morley experiment.

This not only explained, fully and completely, why the Michelson-Morley experiment had failed, but it further shewed that every material measuring rod would necessarily contract just sufficiently to conceal the earth's motion through the ether, so that all similar experiments were doomed to failure in advance. But other types of measuring rods are known to science; beams of light, electric forces, and so on, can be made to span the distances from point to point, and so provide the means for measuring distances. It was thought that where material measuring rods had failed, optical and electrical measuring rods might succeed. The trial was made, repeatedly and in many forms—the names of the late Lord Rayleigh, of Brace and of Trouton are eminent in this connection. And every time it failed. If the earth had a speed x through the ether, every apparatus that the wit of man could devise confused the measurement of x by adding a spurious speed exactly equal to $-x$, and so reiterating the apparent zero answer of the original Michelson-Morley experiment.

The upshot of many years' arduous experimenting was that the forces of nature seemed without exception to be parties to a perfectly organised conspiracy to conceal the earth's motion through the ether. This of course is the language of the layman, not of the man of science. The latter prefers to say that the laws of nature make it impossible to detect the earth's motion through the ether. The philosophical contents of the two statements are precisely identical. In the same way the unscientific inventor may exclaim in despair that the forces of nature are in a conspiracy to prevent his perpetual motion machine from working, while the scientist knows that the obstacle is a far more serious barrier than a conspiracy; it is a natural law. And so, again, the zealous but unenlightened social reformer and the ignorant politician are alike apt to see conspiracies of the deepest dye behind the operation of those economic laws which make it impossible to extract a quart out of a pint pot.

In 1905 Einstein propounded the supposed new law of nature in the form "Nature is such that it is impossible to determine absolute motion by any experiment whatever." It was the first formulation of the principle of relativity.

Oddly enough, it was a reversion to the thought and doctrine of Newton. In his *Principia*, Newton had written:

> It is possible that in the remote regions of the fixed stars or perhaps far beyond them, there may be some body absolutely at rest, but impossible to know, from the positions of bodies to one another in our regions, whether any of these do not keep the same position to that remote body. It follows that absolute rest cannot be determined from the position of bodies in our regions.

He had qualified this by adding:

> I have no regard in this place to a medium, if any such there is, that freely pervades the interstices between the parts of bodies.

In other words, Newton had realised that without an all-pervading ether, it would be impossible to determine the absolute speed of motion through space, and had also seen that such a medium would provide an unmoving standard by reference to which the motions of all bodies could be measured.

The two intervening centuries had seen science busily engaged in discussing the properties of this supposed medium, and now Einstein at one blow deprived it of its most important property of all, that of providing a standard of rest, by reference to which the true speed of any motion could be measured.

Einstein's principle can be stated in another way, which makes its significance stand out more clearly. Astronomy has so far failed to discover Newton's body absolutely at rest, "in the remote regions of the fixed stars, or perhaps far beyond them," so that rest and motion are still merely relative terms. A ship which is becalmed is at rest only in a relative sense—relative to the earth; but the earth is in motion relative to the sun, and the ship with it. If the earth were stayed in its course round the sun, the ship would become at rest relative to the sun, but both would still be moving through the surrounding stars. Check the sun's motion through the stars and there still remains the motion of the whole galactic system of stars relative to the remote nebulae. And these remote nebulae move towards or away from one another with speeds of hundreds of miles a second or more; by going farther into space we not only find no standard of absolute rest, but encounter greater and greater speeds of motion. Unless we have an all-pervading ether to guide us, we cannot even say what we mean by absolute rest, still less can we find it. Einstein's principle now tells us that, so far as all the observable phenomena of nature are concerned, we are free to define "absolute rest" in any way we please.

It is a sensational message. We have a perfect right to say, if we so choose, that this room is at rest, and nature will not say us nay. If the earth has a speed of 1000 miles a second through the ether, then we must suppose that the ether is blowing through this room "like the wind through a grove of trees," at 1000 miles a second. And the principle of relativity assures us that all the phenomena of nature in this room are absolutely unaffected by this 1000 miles-a-second wind, and would indeed be just the same if the wind blew at 100,000 miles a second—or indeed if there were no wind at all.

It is not surprising or even novel that all mechanical phenomena, which have nothing to do with the supposed ether, should be the same; we have seen how this was known to Newton. But if an ether really exists, it seems amazing that the phenomena of optics and of electricity should be the same whether the ether which propagates them is standing still or blowing past and through us at thousands of miles a second. It quite inevitably raises the questions as to whether the ether, whose blowing is supposed to cause the wind, has any existence, or is a mere fiction of our imagination. For we must always remember that the existence of the ether is only an hypothesis, introduced into science by physicists who, taking it for granted that everything must admit of a mechanical explanation, argued that there must be a mechanical medium to transmit waves of light, and all other electrical and magnetic phenomena.

To justify their belief, they had to shew that a system of

pushes, pulls and twists could be devised in the ether to transmit all the phenomena of nature through space and deliver them up at the far end exactly as they are observed —much as a system of bell-wires transmits mechanical force from a bell-pull to a bell. The requisite system of pushes, pulls and twists was found in time, but proved to be exceedingly complicated. Perhaps this was not surprising; the ether had not only to transmit the observed effects, but to conceal its own existence while so doing. It could hardly be a simple matter to arrange that one single mechanism should transmit precisely the same phenomena whether the experimenter sat at rest or dashed through the ether at 1000 miles a second while conducting his experiments. And, in point of fact, the mechanism thus devised proved to be open to the fatal objection that it could only make the two sets of phenomena the same by postulating two distinct mechanisms in these two cases.

We can illustrate the objection by discussing a simple phenomenon in detail. According to this scheme of ethereal transmission, charging a body with electricity sets up a state of strain in the surrounding ether, just like forcing a foreign body into a sea of jelly. When two bodies both at rest in the ether are charged with similar electricity, they repel one another, and their repulsion is supposed to be transmitted through the pressures which this state of strain establishes in the ether.

Suppose, however, that the two charged bodies, instead of being at rest in the ether, are moving through it with precisely the same speed of, say, 1000 miles a second from east to west. As the bodies are still at rest relatively to one another, the principle of relativity shews that the observable phenomena will still be precisely the same as when they were both at absolute rest in the ether. But a quite different mechanism produces the phenomena in this second case. Part of the repulsion is still the result of a strained state of the ether, but not all. The remainder is due to magnetic forces, and these cannot be explained as pressures or tensions in the ether, but have to be attributed to a complicated system of cyclones or whirlwinds.

More complicated electromagnetic phenomena are in general produced by a combination of electric and magnetic forces, and the two kinds of mechanism enter in different proportions with different speeds of motion through the ether. Thus the attempt to find a mechanical explanation of these phenomena involves the need for two distinct mechanisms to produce identically the same phenomenon. It has yet to be shewn that any conceivable ether can accommodate both these mechanisms. But even if this could be proved, such a duality in the mechanism required to produce a single observable phenomenon is so contrary to the usual working of nature that we cannot but feel that we are on the wrong track. Newton's theory of gravitation would have had little chance of acceptance if it had postulated a dual mechanism to explain why an apple fell from a tree, adding that one operated in summer and the other in autumn.

Newton himself laid stress on the necessity for avoiding duplicate mechanisms of this kind. His *Principia* contains a set of "Rules of Reasoning in Philosophy," of which the first two read as follows:

RULE I

We are to admit no more causes of natural things than such as are both true and sufficient to explain their appearances.

To this purpose the philosophers say that Nature does nothing in vain, and more is in vain when less will serve; for Nature is pleased with simplicity, and affects not the pomp of superfluous causes.

RULE II

Therefore to the same natural effects we must, as far as possible, assign the same causes.

As to respiration in a man and in a beast; the descent of stones in Europe and America; the light of our culinary fire and of the sun; the reflection of light in the earth, and in the planets.

There is, however, a stronger case than this against supposing the luminiferous ether to transmit radiation and electrical action.

We have seen how electricity, magnetism, and light all seem to be in a conspiracy to prevent our detecting motion through the ether, but gravitation remains; this has always stood apart from the other phenomena of physics, and has seemed to be of an entirely different nature. Now the law of gravitation involves the idea of distance; it asserts that the gravitational forces between two bodies depend on their distance apart, and so are equal at equal distances. Thus, in theory at least, the law of gravitation provides a measuring-rod for the measurement of distances.

An ether which transmits electrical action can hardly transmit gravitational action as well, since all the properties with which we can endow it are used up in accounting for its transmission of electric and magnetic forces. The measuring-rod which the law of gravitation provides may therefore be expected to be immune from the Fitzgerald-Lorentz contraction, and with such a measuring-rod at our disposal we ought to be able to measure the earth's velocity through space.

Let us examine the possibility in terms of the simplest possible concrete case. Let us idealise our earth, and think of it as a perfect globe. As every point on its surface is now at the same distance from its centre, the force of gravity will be the same at all. If this idealised earth is now set in motion through the ether with a speed of 1000 miles a second, the ordinary Fitzgerald-Lorentz contraction would cause its diameter to shrink by about 600 feet in the direction of motion, and as the points at the end of this contracted diameter are now nearer to the earth's centre than other points on the earth's surface, all movable objects on the earth's surface would tend to slide downhill to these two points.

Even if it existed, this particular effect would be too small to be observed on our actual earth, because the irregularities of mountains and valleys, which we have idealised out of existence, would easily conceal a 600-foot contraction. Yet other gravitational phenomena of a similar kind are large enough to admit of observation, in particular the motions of the perihelia of the planets. And these shew that gravitation is, so to speak, in league with the other forces of nature to conceal motion through the ether; if material measuring-rods experience the Fitzgerald-Lorentz contraction, then the measures of length provided by the law of gravitation do the same. Yet as gravitation cannot be transmitted through the ether, it is hard to see how the measuring-rods of the law of gravitation can be subject to this contraction. We can only conclude that the Fitzgerald-Lorentz contraction does not occur at all, and this compels us to abandon the mechanical ether.

We are compelled to start afresh. Our difficulties have all arisen from our initial assumption that everything in nature, and waves of light in particular, admitted of mechanical explanation: we tried in brief to treat the universe as a huge machine. As this has led us into a wrong path, we must look for some other guiding principle.

A safer guide than the will-o'-the-wisp of mechanical explanations is provided by William of Occam's principle: "Entia non sunt multiplicanda praeter necessitatem." (We must not assume the existence of any entity until we are compelled to do so.) Its philosophical content is identical with that of Newton's first rule of philosophical reasoning quoted above. It is purely destructive; it takes something away, in the present instance the assumption of a mechanical universe with an underlying ether transmitting mechanical action through "empty space," and provides nothing to put in its place.

The obvious way of filling the gap is to introduce the relativity principle: "Nature is such that it is impossible to determine absolute motion by any experiment whatever." At first sight this may seem strange matter with which to fill the void caused by the withdrawal of the ether: the two hypotheses are of such different natures that it may seem incredible that the second should be able to fill the same hole as the first. Yet in actual fact one is almost exactly the antithesis of the other: the primary function of the ether was to provide a fixed frame of reference—all its other properties were ancillaries necessitated by our efforts to reconcile the observed scheme of nature with our preliminary assumption. In its essence, the theory of relativity merely implies the negation of this preliminary assumption, so that the two are exactly antithetical.

Just because this is so, the issue between them is clear cut, and the experiment is capable of deciding it. The verdict is quite unambiguous; we have seen how all experimental efforts to detect an ether have failed, and in so doing have added confirmation to the hypothesis of relativity. Every single experiment ever performed has, so far as we know, decided in favour of the relativity hypothesis.

In this way the hypothesis of a mechanical ether was dethroned, and the principle of relativity set to reign in its stead. The signal for the revolution was a short paper which Einstein published in June 1905. And with its publication, the study of the inner workings of nature passed from the engineer-scientist to the mathematician.

Until this time, we had thought of space as something around us, and of time as something that flowed past us, or even through us. The two seemed to be in every way fundamentally different. We can retrace our steps in space, but never in time; we can move quickly, or slowly, or not at all, in space as we choose, but no one can regulate the rate of flow of time—it rolls on at the same even uncontrollable rate for all of us. Yet Einstein's first results, as interpreted by Minkowski four years later, involved the amazing conclusion that nature knew nothing of all this.

We have already seen how matter is electrical in structure, so that all physical phenomena are ultimately electrical. Minkowski shewed that the theory of relativity required all electrical phenomena to be thought of as occurring, not in space and time separately, as had hitherto been thought, but in space and time welded together so thoroughly that it was impossible to detect any traces of a join, so thoroughly that the whole of the phenomena of nature were unable to divide the product into space and time separately.

When we weld together length and breadth, we get an area—let us say a cricket field. The different players divide it up into its two dimensions in different ways; the direction which is "forwards" for the bowler is "backwards" for the batsman and is left-to-right for the umpire. But the cricket-ball knows nothing of these distinctions: it goes where it is hit, directed only by laws of nature which treat the area of the cricket field as an indivisible whole, length and breadth being welded into a single undifferentiated unit.

If we further weld together an area (such as a cricket field) of two dimensions, and height (of one dimension) we obtain a space of three dimensions. So long as we do this near the earth, we can always call on gravity to separate our space out into "height" and "area"; for instance, the direction of height is that direction in which it is hardest to throw a cricket-ball a given distance. But out in space, nature provides no means of effecting this separation; her laws know nothing of our purely local concepts of horizontal and vertical, and treat space as consisting of three dimensions between which no differentiation is possible.

By a process of welding we have passed in imagination from one dimension to two, and again from two to three. It is harder to pass from three to four because we have no direct experience of a four-dimensional space. And the four-dimensional space which we particularly want to discuss is peculiarly difficult to imagine because one of its dimensions does not consist of ordinary space at all, but of time; to understand the theory of relativity, we are called on to imagine a four-dimensional space in which three dimensions of ordinary space are welded on to one dimension of time.

Let us confront our difficulties singly, by first imagining

a two-dimensional space obtained by welding together one dimension of ordinary space, namely length, and one dimension of time. Fig. 2 may help us to understand the concept. It represents, in diagrammatic form, the running

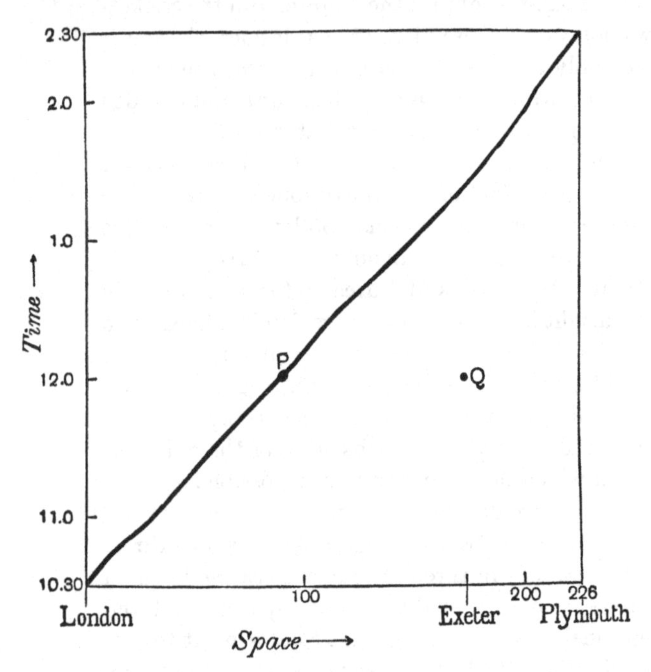

Fig. 2. Diagram to illustrate the motion of a train in space and time

schedule of the Cornish Riviera Express, which leaves Paddington at 10.30 a.m. and reaches Plymouth, 226 miles distant, at 2.30 p.m. The horizontal line represents the 226 miles of track connecting the two stations, and the vertical line represents the interval of time from 10.30 a.m. to 2.30 p.m. on any day on which the train is running.

The thick line represents the progress of the train. For instance the point P on this line is opposite the time 12.0 noon, and above the distance $91\frac{1}{2}$ miles from Paddington, indicating that the train has travelled $91\frac{1}{2}$ miles by noon. On the other hand a point such as Q represents a spot somewhere near Exeter at noon; it does not lie on the thick line, because the train does not reach Exeter by noon. The whole area of the diagram represents all possible spots on the line between Paddington and Plymouth at all times between 10.30 a.m. and 2.30 p.m. Thus by welding together a length, namely 226 miles of track, and a time, namely four hours around mid-day, we have obtained an area having one dimension of space and one of time.

In the same way we can imagine the three dimensions of space and one dimension of time welded together, forming a four-dimensional volume which we shall describe as a "continuum." Then the principle of relativity, as interpreted by Minkowski, states that all the phenomena of electromagnetism may be thought of as occurring in a continuum of four dimensions—three dimensions of space and one of time—*in which it is impossible to separate the space from the time in any absolute manner.* In other words the continuum is one in which space and time are so completely welded together, so perfectly merged into one, that the laws of nature make no distinction between them, just as, on the cricket field, length and breadth are so perfectly merged into one that the flying cricket-ball makes no distinction between them, treating the field merely as an area in which length and breadth separately have lost all meaning.

It may be objected that Fig. 2 gives no help towards imagining this continuum; that it is merely diagrammatic; that it does not really represent the welding together of true time and length, but merely of one length with another length, which as everyone knows gives an area—in this case a page of the book. We need not linger over this objection because our final conclusion will be that the four-dimensional continuum is, in much the same sense, also purely diagrammatic. It merely provides a convenient framework in which to exhibit the workings of nature, just as Fig. 2 provides a convenient framework in which to exhibit the running of a train.

Yet, just because we can exhibit all nature within this framework, it must correspond to some sort of an objective reality. But its division into space and time is not objective; it is merely subjective. If you and I happen to be moving with different speeds, space and time mean something different to you from what they mean to me; we divide the continuum into space and time in different ways, just as, if we happen to be facing in different directions, "in front" and "to the left" have different meanings for the two of us, or just as the bowler and the batsman divide up a cricket field in different ways of which the cricket-ball knows nothing. Even if I change my own speed of motion, by putting on the brakes of my car, or jumping on to a moving bus, I am re-arranging the division of the continuum into space and time for myself. And the essence of the theory of relativity is that nature knows nothing of these divisions of the continuum into space and time; in Minkowski's words: "Space and time separately have vanished into the merest shadows, and only a sort of combination of the two preserves any reality."

This shews in a flash why the old luminiferous ether had inevitably to fade out of the picture—it claimed to fill "all space," and so to divide up the continuum objectively into time and space. And the laws of nature, not recognising such divisions as a possibility, cannot recognise the existence of the ether as a possibility.

Thus if we want to visualise the propagation of light-waves and electromagnetic forces by thinking of them as disturbances in an ether, our ether must be something very different from the mechanical ether of Maxwell and Faraday. It may be thought of as a four-dimensional structure, filling up the whole continuum, and so extending through all space and all time, in which case we can all enjoy the same ether. Or, if we want a three-dimensional ether, it must be subjective in a way in which the Maxwell-Faraday ether was not. Each of us must then carry his own ether about with him, much as in a shower of rain each observer carries his own rainbow about with him. If I change my speed of motion I create a new ether for myself, just as, if I step a few paces in a sunny shower, I acquire a new rainbow for myself. And unless the expanding universe described above (p. 56) is a pure illusion, everyone's ether must incessantly expand and stretch. Whether a structure of this kind ought to be called an ether, is open to question; it would be hard to find any property it has in common with the old nineteenth-century ether. Indeed, as the hypothesis of relativity is the exact negation of the existence of the old ether, it is clear that any ether that relativity can allow to remain in being must be the exact opposite of the old ether. This being so, it seems a mistaken effort to call them by the same name.

I do not think there is any real divergence of opinion

among competent scientists on all this. Sir Arthur Eddington truly says that about half the leading physicists assert that the ether exists and the other half deny its existence, but continues: "Both parties mean exactly the same thing, and are divided only by words." Sir Oliver Lodge, who has been the staunchest supporter of the objective existence of an ether in recent years, writes:

> The ether in its various forms of energy dominates modern physics, though many prefer to avoid the term "ether" because of its nineteenth-century associations, and use the term "space." The term used does not much matter.

Clearly, if it is a matter of indifference whether we speak of the ether or of space, of the existence or non-existence of the ether, then even its most ardent devotees cannot claim much objective reality for it. I think the best way of regarding the ether is as a frame of reference just as the diagram on p. 88 is a frame of reference; its existence is just as real, and just as unreal, as that of the equator, or the north pole, or the meridian of Greenwich. It is a creation of thought, not of solid substance. We have seen how *the* ether, which is the same for all of us, as distinguished from your ether or my ether, must be supposed to pervade all time as well as all space, and that no valid distinction can be drawn between its occupancy of time and space. The framework in time to which we must compare the time-dimension of the ether is of course ready to hand—it is the division of the day into hours, minutes and seconds. And unless we think of this division as material, which no one ever does or has done, we are not justified in thinking of the ether as material. In the new light which the theory of relativity has cast over science, we see that a material ether filling space could only be accompanied by a material ether filling time—the two stand or fall together.

Thus we seem on fairly safe ground in thinking of the ether as a pure abstraction; it is at best "a local habitation and a name." Yet a local habitation for what? The universe consists only of waves, and we first introduced the ether as the nominative of the verb "to undulate." This conception must now be abandoned, for the utterly unsubstantial ether we are now considering is as incapable of undulation as is the equator or the meridian of Greenwich. It does not of course follow that nothing undulatory can be propagated through this immaterial medium. We speak of a heat-wave, or a suicide-wave, and do not ask for an undulating medium to convey them. The heat-wave might be propagated round the equator, and the suicide-wave along the meridian of Greenwich.

It may be thought that, although we can obtain no direct evidence of the existence of the ether, yet we can find evidence of something of the nature of waves passing through it, in all the phenomena which are generally taken to prove the undulatory nature of light—Newton's rings, diffraction patterns, and interference phenomena in general. This, however, is not so, for again we have no knowledge of the supposed waves except where there are particles of matter to reveal them to us. The phenomena just mentioned give us no knowledge of things passing through the ether, but only of things falling on matter. So far as we know, nothing at all is propagated that is more concrete than a mathematical abstraction—it is like astronomical noon being propagated over the surface of the earth as the earth turns round under the sun. Yet I can imagine a physicist intervening with an objection at this stage; it would be something like this:

Physicist. The sunshine out of doors represents energy which was generated in the sun. Eight minutes ago it was in the sun; now it is here. Consequently it must have come from the sun, and so must have travelled through the space which intervenes between the sun and us. It seems to me, then, that energy must be propagated through space.

Mathematician. Let us make the question at issue as precise as possible. Let us fix our attention on a definite parcel of sunlight, say that which falls on my book in the space of a second, as I sit reading out in the bright sunshine. This, you say, was in the sun eight minutes ago. Four minutes ago it was, I suppose, out in space, half-way between the sun and ourselves. Two minutes ago it was three-quarters of the way towards us?

Physicist. Yes; and that is what I call being propagated through space; energy moves from one bit of space to another.

Mathematician. Your concept implies that at any instant the different little bits of space are occupied by different amounts of energy. If so, it ought of course to be possible to calculate or measure how much is in a given bit of space at a given instant. If you assume that the sun is at rest in an ether, and that sunlight is energy propagated through this ether, then, I admit, you can get a quite definite answer to the problem; Maxwell gave it in 1863. Also if you assume that the sun, and of course the whole solar system with it, is moving steadily through the ether at a known speed, say 1000 miles a second, you can also get a definite answer to your problem. But—and this is the crux of the matter—the two answers are different. Will you tell me which is the right one?

Physicist. Obviously the first is right if the sun is at rest in the ether, and the second if the sun has a steady speed of 1000 miles a second through the ether.

Mathematician. Yes, but we are in agreement that "at rest in the ether" means nothing at all, and "a steady speed of 1000 miles a second through the ether" means nothing at all. If we try to attach any meaning to them, all the phenomena of nature insist that the same meaning must be attached to both. Consequently I find your answer meaningless.

In some such way as this we find that the attempt to parcel out energy amongst the different parts of space leads to an ambiguity which cannot be resolved. It seems natural to suppose that our attempt is a misguided one, and that the partition of energy through space is illusory.

And again, the attempt to regard the flow of energy as a concrete stream always defeats itself. With a stream of water, we can say that a certain particle of water is now here, now there; with energy it is not so. The concept of energy flowing about through space is useful as a picture, but leads to absurdities and contradictions if we treat it as a reality. Professor Poynting gave a well-known formula which tells us how energy may be pictured as flowing in a certain way, but the picture is far too artificial to be treated as a reality; for instance, if an ordinary bar-magnet is electrified and left standing at rest, the formula pictures energy flowing endlessly round and round the magnet, rather like innumerable rings of children joining hands and dancing to all eternity round a maypole. The mathematician brings the whole problem back to reality by treating this flow of energy as a mere mathematical abstraction. Indeed he is almost compelled to go farther and treat energy itself as a mere mathematical abstraction—the constant of integration in a differential equation. If he does this, it becomes no more absurd that there should be two different values for the amounts of energy in a given region of space than that there should be two different times at the same place, such as standard and daylight-saving times in New York, or civil and sidereal times in an observatory. If he declines to do this, he is left to defend the untenable position that the universe is built, in a concrete way, of energy in its alternative forms of matter and radiation, and that energy cannot be localised in space. We shall discuss this situation further below (p. 129).

Before proceeding to consider other developments of the theory of relativity, it seems appropriate to discard the word "ether" in favour of the term "continuum," this meaning the four-dimensional "space" we have already imagined, in which the three dimensions of ordinary space are supplemented by time acting as a fourth dimension.

Laws of nature express happenings in time and space, and so can of course be stated with reference to this four-dimensional continuum. In discussing these laws quantitatively, it is found convenient to imagine both time and space measured in a very special and a very artificial manner. We shall not measure lengths in terms of feet or centimetres, but in terms of a unit of about 186,000 miles, which is the distance that light travels in a second. And we shall not measure time in ordinary seconds, but in terms of a mysterious unit equal to a second multiplied by $\sqrt{-1}$ (the square root of -1). Mathematicians speak of $\sqrt{-1}$ as an "imaginary" number, because it has no existence outside their imaginations, so that we are measuring time in a highly artificial manner. If we are asked why we adopt these weird methods of measurement, the answer is that they appear to be nature's own system of measurement; at any rate they enable us to express the results of the theory of relativity in the simplest possible form. If we are further asked why this is so, we can give no answer—if we could, we should see far deeper than we now do into the inner mysteries of nature.

Let us, then, agree to use the weird system of measurement just described, and construct our continuum accordingly. Minkowski shewed that if the hypothesis of relativity is true, the statement of the laws of nature must shew no distinction between time and space, when the continuum is constructed in the way just described; the three dimensions of space and one of time enter as absolutely equal partners into the formulation of every natural law. If they did not, the law would be at variance with the principle of relativity.

It was soon noticed that Newton's famous law of gravitation did not conform to the condition just stated, so that either Newton's law or the hypothesis of relativity was wrong. Einstein examined what alterations would have to be applied to Newton's law to bring it into conformity with the hypothesis of relativity, and found that the necessary changes involved the appearance of three new phenomena which were not implied in Newton's old law. In other words, nature provided three distinct ways of deciding observationally between the laws of Einstein and Newton. When the test was made, the decision was favourable to Einstein in every case.

What we call the "law of gravitation" is, strictly speaking, nothing more than a mathematical formula giving the acceleration of a moving body—the rate at which it changes its speed of motion. Newton's law lent itself to a rather obvious mechanical interpretation: a body moved in the same way as it would if it were "drawn off from its rectilinear motion" (to use Newton's phrase) by a force proportional to the inverse square of the distance. Newton accordingly supposed such a force to exist; it was called the "force of gravity." Einstein's law did not lend itself to any such interpretation in terms of forces, or indeed to any mechanical interpretation whatever—still another indication, if one were needed, that the age of mechanical science had passed. But it was found to admit of an easy interpretation in terms of geometry. The effect of a mass of gravitating matter was not, as Newton had imagined, to exude a "force," but to distort the four-dimensional continuum in its neighbourhood. The moving planet or cricket-ball was no longer drawn off from its rectilinear motion by the pull of a force, but by a curvature of the continuum.

It is difficult enough to imagine the four-dimensional continuum even when undistorted, and still more so to imagine its distortions, but the two-dimensional analogy of an area may help. Surfaces such as a cricket field or the skin of our hand are two-dimensional continua; the analogies of the distortions produced by gravitating masses are mole-hills or blisters. The cricket-ball which rolls over a mole-hill is "drawn off from its rectilinear motion" like a comet or a ray of light passing near the sun. And the combined distortions of the four-dimensional continuum produced by all the matter in the universe cause the continuum to bend back on itself to form a closed surface, so that space becomes "finite," with the results that have been already discussed in Chapter II.

Space and time as separate entities have already disappeared from the universe; gravitational forces now disappear also, leaving nothing but a crumpled continuum.

Nineteenth-century science had reduced the universe to a playground of forces of only two kinds—gravitational forces which govern the major phenomena of astronomy, besides keeping our bodies and possessions on the earth's surface, and electromagnetic forces, which control all other physical phenomena, such as light, heat, sound, cohesion, elasticity, chemical change, and so forth. Now that gravitational forces have disappeared from science, it is natural to wonder why electromagnetic forces happen to survive, and how they figure in the continuum. Although the question is not finally settled, it seems likely that these, too, are destined to go the way of gravitational forces. Weyl and Eddington successively propounded theories which dispensed with electromagnetic forces altogether, and tried to interpret all physical phenomena as consequences of the peculiar geometry of the continuum. Both these proved open to objections; the fate of a more recent theory of the same type by Einstein is still in the balance. But whatever theory finally prevails, it seems fairly certain that in some way or other electromagnetic forces will ere long be resolved merely into a new type of crumpling of the continuum, essentially different in its geometry, but in no other respect, from that whose effects we describe as gravitation. If so, the universe will have resolved itself into an empty four-dimensional space, totally devoid of substance, and totally featureless except for the crumplings, some large and some small, some intense and some feeble, in the configuration of the space itself.

What we have hitherto spoken of as the propagation of energy, such as the passage of sunlight from sun to earth, now reduces to nothing more than the continuity of a corrugated crumpling along a line in the continuum which extends over about eight minutes of our terrestrial time and about 92,500,000 miles of our terrestrial length. We now see that we cannot picture it as the propagation of anything concrete or objective through space unless we first divide the continuum objectively into space and time, and this is precisely what we are forbidden to do.

To sum up, a soap-bubble with irregularities and corrugations on its surface is perhaps the best representation, in terms of simple and familiar materials, of the new universe revealed to us by the theory of relativity. The universe is not the interior of the soap-bubble but its surface, and we must always remember that, while the surface of the soap-bubble has only two dimensions, the universe-bubble has four—three dimensions of space and one of time. And the substance out of which this bubble is blown, the soap-film, is empty space welded on to empty time.

Chapter V

INTO THE DEEP WATERS

Let us study in more detail this soap-bubble, blown of emptiness, by which modern science portrays the universe. Its surface is richly marked with irregularities and corrugations. Two main kinds may be discerned, which we interpret as radiation and matter, the ingredients of which the universe appears to us to be built.

Markings of the first kind represent radiation. All radiation travels at the same uniform speed of about 186,000 miles a second. If the train in Fig. 2 (p. 88) had travelled at a uniform speed of a mile a minute, its motion would have been represented by a perfectly straight line inclined at an angle of 45° to the vertical. A succession of trains all moving uniformly at a mile a minute would be represented by a lot of lines all parallel to this. Now let us change our standard speed from a mile a minute to 186,000 miles a second, and replace the one direction from London to Plymouth by all the directions in space. The diagram on p. 88 now becomes replaced by the four-dimensional continuum, and radiation is represented by a set of lines all making the same angle (45°) with the direction of time advancing.

Markings of the second kind represent matter. This moves through space at a variety of different speeds, but all are small in comparison with the speed of light. To a first rough approximation, we may regard all matter as standing still in space, and moving forward only in time, so that the markings which represent it run in the direction of time advancing, just as, if the train whose journey is shewn in Fig. 2 (p. 88) were to stop at a station, its stay there would be represented by a bit of vertical line.

The markings which represent matter tend to form broad bands across the surface of the soap-bubble, like broad streaks of paint on a canvas. This is because the matter of the universe tends to aggregate into large masses —stars and other astronomical bodies. These bands or streaks are known as "world lines"; the world line of the sun traces out the position of the sun in space which corresponds to each instant of time. We can picture this diagrammatically in Fig. 3, opposite.

Just as a cable is formed of a great number of fine threads, so the world line of a large body like the sun is formed of innumerable smaller world lines, the world lines of the separate atoms of which the sun is composed. Here and there these fine threads enter or leave the main cable as an atom is swallowed up by, or ejected from, the sun.

We may think of the surface of the bubble as a tapestry whose threads are the world lines of atoms. In so far as atoms are permanent and indestructible, the thread-like world lines of the atoms traverse the whole length of the picture in the direction of time advancing. But if atoms are annihilated, the threads may end abruptly and tassels of world lines of radiation spread out from their broken ends. As we move timewards along the tapestry, its various threads for ever shift about in space and so change their places relative to one another. The loom has been set so that they are compelled to do this according to definite rules which we call the "laws of nature."

The world line of the earth is a smaller cable, made up of

several strands, these representing the mountains, trees, aeroplanes, human bodies and so on, the aggregate of which makes up the earth. Each strand is made up of many

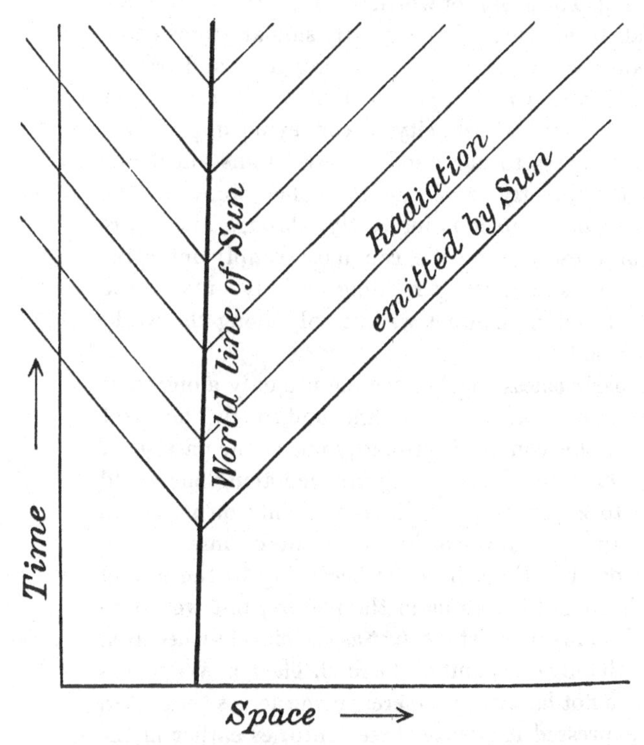

Fig. 8. Diagram to illustrate the motion of the sun and its radiation in space and time (cf. Fig. 2 on p. 88)

threads—the world lines of its atoms. A strand which represents a human body does not differ in any observable essentials from the other strands. It shifts about, relatively to the other strands, less freely than an aeroplane, but more freely than a tree. Like the tree, it begins as a small thing and increases by continual absorption of atoms from outside—its food. The atoms of which it is formed do not differ in essentials from other atoms; exactly similar atoms enter into the composition of mountains, aeroplanes and trees.

Yet the threads which represent the atoms of a human body have the special capacity of conveying impressions through our senses to our minds. These atoms affect our consciousness directly, while all the other atoms of the universe can only affect it indirectly, through the intermediary of these atoms. We can most simply interpret consciousness as something residing entirely outside the picture, and making contact with it only along the world lines of our bodies.

Your consciousness touches the picture only along your world line, mine along my world line, and so on. The effect produced by this contact is primarily one of the passage of time; we feel as if we were being dragged along our wörld line so as to experience the different points on it, which represent our states at the different instants of time, in turn.

It may be that time, from its beginning to the end of eternity, is spread before us in the picture, but we are in contact with only one instant, just as the bicycle-wheel is in contact with only one point of the road. Then, as Weyl puts it, events do not happen; we merely come across them. Or, as Plato expressed it twenty-three centuries earlier in the *Timaeus*:

> The past and future are created species of time which we unconsciously but wrongly transfer to the eternal essence. We say "was," "is," "will be," but the truth is that "is" can alone properly be used.

In this case, our consciousness is like that of a fly caught in a dusting-mop which is being drawn over the surface of the picture; the whole picture is there, but the fly can only experience the one instant of time with which it is in immediate contact, although it may remember a bit of the picture just behind it, and may even delude itself into imagining it is helping to paint those parts of the picture which lie in front of it.

Or again, it may be that our consciousness should be compared to the feeling in the finger of the painter as he guides the brush forward over the still unfinished picture. If so, the impression of influencing the parts of the picture yet to come is something more than a pure illusion. At present science can tell us very little as to the way in which our consciousness apprehends the picture; it is concerned mainly with the nature of the picture.

We have seen how the ether which was at one time supposed to fill the universe has been reduced to an abstraction, a framework of empty space, amounting to nothing more than the spatial dimensions of a soap-bubble, whose soap-film consists of vacancy. The waves which were at one time supposed to traverse this ether have also been reduced to little more than an abstraction: they are corrugations on a cross-section of the bubble by time.

This quality of abstractness in what were at one time regarded as material "ether-waves" recurs in a far more acute form when we turn to the system of waves which make up an electron. The "ether" in terms of which we find it convenient to explain ordinary radiation—say, sunlight—has three dimensions of space, in addition to its one dimension of time. So also has the ether in which we describe the waves which constitute a single electron isolated in space; this may or may not be the same ether as before, but it is similar in having three dimensions of space and one of time. But a single electron isolated in space provides a perfectly eventless universe, the simplest conceivable event occurring when two electrons meet one another. And to describe, in its simplest terms, what happens when two electrons meet one another, the wave-mechanics asks for a system of waves in an ether which has seven dimensions; six are of space, three for each of the electrons, and one is of time. To describe a meeting of three electrons, we need an ether of ten dimensions, nine of space (again three for each electron) and one of time. Were it not for the last dimension of time which binds all the others together, the various electrons would all exist in separate non-communicating three-dimensional spaces. Thus time figures as the mortar which binds the bricks of matter together, much as, on the spiritual plane, the "windowless monads" of Leibnitz were bound together by the universal mind. Or, perhaps with a nearer approach to actuality, we may think of the electrons as objects of thought, and time as the process of thinking.

Most physicists would, I think, agree that the seven-dimensional space in which the wave-mechanics pictures the meeting of two electrons is purely fictitious, in which case the waves which accompany the electrons must also be regarded as fictitious. Thus Professor Schrödinger, writing of the seven-dimensional space, says that although it

> has quite a definite physical meaning, it cannot very well be said to "exist"; hence a wave-motion in this space cannot be said to "exist" in the ordinary sense of the word either. It is merely an adequate mathematical description of what happens. It may be that also in the case of one single [electron], the wave-

motion must not be taken to "exist" in *too* literal a sense, although the configuration-space happens to coincide with ordinary space in this particularly simple case.

Yet it is hard to see how we can attribute a lower degree of reality to the one set of waves than to the other: it is absurd to say that the waves of single electrons are real, while those of pairs of electrons are fictitious. And the waves of single electrons are real enough to record themselves on a photographic plate and produce the patterns shewn in Plate II. We can only regain complete consistency by supposing all the waves, those of two electrons, those of one electron, and the waves on Professor Thomson's photographic plate, to have the same degree of reality or unreality.

Some physicists meet this situation by regarding the electron-waves as waves of probability. When we speak of a tidal-wave we mean a material wave of water which wets everything in its path. When we speak of a heat-wave we mean something which, although not material, warms up everything in its path. But when the evening papers speak of a suicide-wave, they do not mean that each person in the path of the wave will commit suicide; they merely mean that the likelihood of his doing so is increased. If a suicide-wave passes over London, the death-rate from suicide goes up; if it passes over Robinson Crusoe's island, the probability that the sole inhabitant will kill himself goes up. The waves which represent an electron in the wave-mechanics may, it is suggested, be probability-waves, whose intensity at any point measures the probability of the electron being at that point.

Thus at each point on Professor Thomson's plate (Figs. 2 and 3, Plate II), the wave-intensity measures the probability that a single diffracted electron would hit the plate at that spot. When a whole crowd of electrons is diffracted, the total number which hit any spot is of course proportional to the probability of each individual hitting the spot, so that the darkening of the plate gives a measure of the probability per electron.

This view has the great merit that it enables the electrons to preserve their identity. If the electron-waves were true material waves, each system of waves would probably be dispersed by the experiment, so that no electrified particles would survive as such in the diffracted beam. Indeed, any encounter with matter would break up electrons, which could not be regarded as permanent structures. Actually of course it is the shower of electrons, rather than the individual, that is diffracted; the individual electrons move as particles and retain their identity as such.

All this is in accordance with Heisenberg's "uncertainty principle" (p. 22), which makes it impossible ever to say: "An electron is here, at this precise spot, and is moving at just so many miles an hour"; it is also in accordance with the general principle of Dirac, which has already been explained (p. 25). Yet these two principles alone are not enough to specify the full nature of the electron-waves.

Heisenberg and Bohr have suggested that these waves must be regarded merely as a sort of symbolic representation of our knowledge as to the probable state and position of an electron. If so, they change as our knowledge changes, and so become largely subjective. Thus we need hardly think of the waves as being located in space and time at all; they are mere visualisations of a mathematical formula of an undulatory, but wholly abstract, nature.

A still more drastic possibility, again arising out of a

suggestion made by Bohr, is that the minutest phenomena of nature do not admit of representation in the space-time framework at all. On this view the four-dimensional continuum of the theory of relativity is adequate only for some of the phenomena of nature, these including large-scale phenomena and radiation in free space; other phenomena can only be represented by going outside the continuum. We have, for instance, already tentatively pictured consciousness as something outside the continuum, and have seen how the meeting of two electrons can most simply be pictured in seven dimensions. It is conceivable that happenings entirely outside the continuum determine what we describe as the "course of events" inside the continuum, and that the apparent indeterminacy of nature may arise merely from our trying to force happenings which occur in many dimensions into a smaller number of dimensions. Imagine, for instance, a race of blind worms, whose perceptions were limited to the two-dimensional surface of the earth. Now and then spots of the earth would sporadically become wet. We, whose faculties range through three dimensions of space, call the phenomenon a rain-shower, and know that events in the third dimension of space determine, absolutely and uniquely, which spots shall become wet and which shall remain dry. But if the worms, unconscious even of the existence of the third dimension of space, tried to thrust all nature into their two-dimensional framework, they would be unable to discover any determinism in the distribution of wet and dry spots; the worm-scientists would only be able to discuss the wetness and dryness of minute areas in terms of probabilities, which they would be tempted to treat as ultimate truth. Although the time is not yet ripe for a decision, this seems to me, personally, the most promising interpretation of the situation. Just as the shadows on a wall form the projection of a three-dimensional reality into two dimensions, so the phenomena of the space-time continuum may be four-dimensional projections of realities which occupy more than four dimensions, so that events in time and space become

> no other than a moving row
> of Magic Shadow-shapes that come and go.

It may perhaps be objected that we have paid altogether too much attention to the wave-mechanics, which after all is only a mathematical picture, when probably innumerable other mathematical pictures might serve equally well, and might lead to entirely different conclusions.

It is true that the wave-mechanics picture can make no claim to uniqueness. Other systems are in the field, particularly those of Heisenberg and Dirac. Yet in the main these only say the same thing in other, and frequently more complicated, words. No other system yet devised explains things so simply, or seems to be so true to nature, as the wave-mechanics of de Broglie and Schrödinger. Photographs such as those shewn in Plate II bear witness that waves of definite wave-length are somehow fundamental in nature's scheme; these waves form the fundamental concept of the wave-mechanics, but only appear as rather far-fetched by-products in the other systems. Also, just because of its inherent simplicity, the wave-mechanics has shewn a capacity for penetrating much farther into the secrets of nature than any other system, so that other systems are already falling somewhat into the background. To vary our metaphor, they have served a valuable purpose as scaffolding, but there seems to be but little inclination to add to them further.

If then we are to concentrate on one picture, we seem justified in selecting that provided by the wave-mechanics, although in point of fact either the system of Heisenberg or that of Dirac would lead us to very much the same conclusion. The essential fact is simply that *all* the pictures which science now draws of nature, and which alone seem capable of according with observational fact, are *mathematical* pictures.

Most scientists would agree that they are nothing more than pictures—fictions if you like, if by fiction you mean that science is not yet in contact with ultimate reality. Many would hold that, from the broad philosophical standpoint, the outstanding achievement of twentieth-century physics is not the theory of relativity with its welding together of space and time, or the theory of quanta with its present apparent negation of the laws of causation, or the dissection of the atom with the resultant discovery that things are not what they seem; it is the general recognition that we are not yet in contact with ultimate reality. To speak in terms of Plato's well-known simile, we are still imprisoned in our cave, with our backs to the light, and can only watch the shadows on the wall. At present the only task immediately before science is to study these shadows, to classify them and explain them in the simplest possible way. And what we are finding, in a whole torrent of surprising new knowledge, is that the way which explains them more clearly, more fully and more naturally than any other is the mathematical way, the explanation in terms of mathematical concepts. It is true, in a sense somewhat different from that intended by Galileo, that "Nature's great book is written in mathematical language." So true is it that no one except a mathematician need ever hope fully to understand those branches of science which try to unravel the fundamental nature of the universe—the theory of relativity, the theory of quanta and the wave-mechanics.

The shadows which reality throws onto the wall of our cave might *à priori* have been of many kinds. They might conceivably have been perfectly meaningless to us, as meaningless as a cinematograph film shewing the growth of microscopic tissues would be to a dog who had strayed into a lecture-room by mistake. Indeed our earth is so infinitesimal in comparison with the whole universe, we, the only thinking beings, so far as we know, in the whole of space, are to all appearances so accidental, so far removed from the main scheme of the universe, that it is *à priori* all too probable that any meaning that the universe as a whole may have, would entirely transcend our terrestrial experience, and so be totally unintelligible to us. In this event, we should have had no foothold from which to start our exploration of the true meaning of the universe.

Although this is the most likely event, it is not impossible that some of the shadows thrown onto the walls of our cave might suggest objects and operations with which we cave-dwellers were already familiar in our caves. The shadow of a falling body behaves like a falling body, and so would remind us of bodies we had ourselves let fall; we should be tempted to interpret such shadows in mechanical terms. This explains the mechanical physics of the last century; the shadows reminded our scientific predecessors of the behaviour of jellies, spinning-tops, thrust-bars, and cog-wheels, so that they, mistaking the shadow for the substance, believed they saw before them a universe of jellies and mechanical devices. We know now that the interpretation is conspicuously inadequate: it fails to explain the simplest phenomena, the propagation of a sunbeam, the composition of radiation, the fall of an apple, or the whirl of electrons in the atom.

Again, the shadow of a game of chess, played by the actors out in the sunlight, would remind us of the games of chess we had played in our cave. Now and then we might recognise knights' moves, or observe castles moving simultaneously with kings and queens, or discern other characteristic moves so similar to those we were accustomed to play that they could not be attributed to chance. We would no longer think of the external reality as a machine; the details of its operation might be mechanical, but in essence it would be a reality of thought: we should recognise the chess players out in the sunlight as beings governed by minds like our own; we should find the counterpart of our own thoughts in the reality which was for ever inaccessible to our direct observation.

And when scientists study the world of phenomena, the shadows which nature throws onto the wall of our cave, they do not find these shadows totally unintelligible, and neither do they seem to represent unknown or unfamiliar objects. Rather, it seems to me, we can recognise chess-players outside in the sunshine who appear to be very well acquainted with the rules of the game *as we have formulated them in our cave*. To drop our metaphor, nature seems very conversant with the rules of pure mathematics, as our mathematicians have formulated them in their studies, out of their own inner consciousness and without drawing to any appreciable extent on their experience of the outer world. By "pure mathematics" is meant those departments of mathematics which are creations of pure thought, of reason operating solely within her own sphere, as contrasted with "applied mathematics" which reasons about the external world, after first taking some supposed property of the external world as its raw material. Descartes, looking round for an example of the produce of pure thought uncontaminated by observation (rationalism), chose the fact that the sum of the three angles of a triangle was necessarily equal to two right angles. It was, as we now know, a singularly unfortunate choice. Other choices, far less open to objection, might easily have been made, as, for instance, the laws of probability, the rules of manipulation of "imaginary" numbers—i.e. numbers containing the square roots of negative quantities—or multi-dimensional geometry. All these branches of mathematics were originally worked out by the mathematician in terms of abstract thought, practically uninfluenced by contact with the outer world, and drawing nothing from experience: they formed

an independent world
created out of pure intelligence.

And now it emerges that the shadow-play which we describe as the fall of an apple to the ground, the ebb and flow of the tides, the motion of electrons in the atom, are produced by actors who seem very conversant with these purely mathematical concepts—with our rules of our game of chess, which we formulated long before we discovered that the shadows on the wall were also playing chess.

When we try to discover the nature of the reality behind the shadows, we are confronted with the fact that all discussion of the ultimate nature of things must necessarily be barren unless we have some extraneous standards against which to compare them. For this reason, to borrow Locke's phrase, "the real essence of substances" is for ever unknowable. We can only progress by discussing the laws which govern the changes of substances, and so produce the phenomena of the external world. These we can compare with the abstract creations of our own minds.

For instance, a deaf engineer studying the action of a pianola might try first to interpret it as a machine, but would be baffled by the continuous reiteration of the intervals 1, 5, 8, 13 in the motions of its trackers. A deaf musician, although he could hear nothing, would immediately recognise this succession of numbers as the intervals of the common chord, while other successions of less frequent occurrence would suggest other musical chords. In this way he would recognise a kinship between his own thoughts and the thoughts which had resulted in the making of the pianola; he would say that it had come into existence through the thought of a musician. In the same way, a scientific study of the action of the universe has suggested a conclusion which may be summed up, though very crudely and quite inadequately, because we have no language at our command except that derived from our terrestrial concepts and experiences, in the statement that the universe appears to have been designed by a pure mathematician.

This statement can hardly hope to escape challenge on the ground that we are merely moulding nature to our preconceived ideas. The musician, it will be said, may be so engrossed in music that he would contrive to interpret every piece of mechanism as a musical instrument; the habit of thinking of all intervals as musical intervals may be so ingrained in him that if he fell downstairs and bumped on stairs numbered 1, 5, 8 and 13 he would see music in his fall. In the same way, a cubist painter can see nothing but cubes in the indescribable richness of nature—

and the unreality of his pictures shews how far he is from understanding nature; his cubist spectacles are mere blinkers which prevent his seeing more than a minute fraction of the great world around him. So, it may be suggested, the mathematician only sees nature through the mathematical blinkers he has fashioned for himself. We may be reminded that Kant, discussing the various modes of perception by which the human mind apprehends nature, concluded that it is specially prone to see nature through mathematical spectacles. Just as a man wearing blue spectacles would see only a blue world, so Kant thought that, with our mental bias, we tend to see only a mathematical world. Does our argument merely exemplify this old pitfall, if such it is?

A moment's reflection will shew that this can hardly be the whole story. The new mathematical interpretation of nature cannot all be in our spectacles—in our subjective way of regarding the external world—since if it were we should have seen it long ago. The human mind was the same in quality and mode of action a century ago as now; the recent great change in scientific outlook has resulted from a vast advance in scientific knowledge and not from any change in the human mind; we have found something new and hitherto unknown in the objective universe outside ourselves. Our remote ancestors tried to interpret nature in terms of anthropomorphic concepts of their own creation and failed. The efforts of our nearer ancestors to interpret nature on engineering lines proved equally inadequate. Nature refused to accommodate herself to either of these man-made moulds. On the other hand, our efforts to interpret nature in terms of the concepts of pure mathematics have, so far, proved brilliantly successful. It would now seem to be beyond dispute that in some way nature is more closely allied to the concepts of pure mathematics than to those of biology or of engineering, and even if the mathematical interpretation is only a third man-made mould, it at least fits objective nature incomparably better than the two previously tried.

A hundred years ago, when scientists were trying to interpret the world mechanically, no wise man came forward to assure them that the mechanical view was bound to prove a misfit in the end—that the phenomenal universe would never make sense until it was projected on to a screen of pure mathematics: had they brought forward a convincing argument to this effect, science might have been saved much fruitless labour. If the philosopher now says—"What you have found is nothing new: I could have told you that it must be so all the time," the scientist may reasonably inquire—"Why, then, did you not tell us so, when we should have found the information of real value?"

Our contention is that the universe now appears to be mathematical in a sense different from any which Kant contemplated or possibly could have contemplated—in brief, the mathematics enters the universe from above instead of from below.

In one sense it may be argued that everything is mathematical. The simplest form of mathematics is arithmetic, the science of numbers and quantities—and these permeate the whole of life. For instance, commerce, which consists largely of the arithmetical operations of book-keeping, stock-taking and so on, is in a sense a mathematical occupation—but it is not in this sense that the universe now appears to be mathematical.

Again, every engineer has to be something of a mathematician; if he is to calculate and predict the mechanical behaviour of bodies with accuracy, he must use mathematical knowledge and look at his problems through mathematical spectacles—but again it is not in this way that science has begun to see the universe as mathematical. The mathematics of the engineer differs from the mathematics of the shopkeeper only in being far more complex. It is still a mere tool for calculation; instead of evaluating stock-in-trade or profits, it evaluates stresses and strains or electric currents.

On the other hand Plutarch records that Plato used to say that God for ever geometrises—*Πλάτων ἔλεγε τὸν θεὸν ἀεὶ γεωμετρεῖν*—and he sets an imaginary symposium at work to discuss what Plato meant by this. Clearly he meant something quite different in kind from what we mean when we say that the banker for ever arithmetises. Among the illustrations given by Plutarch are: that Plato had said that geometry sets limits to what would otherwise be unlimited, and that he had stated that God had constructed the universe on the basis of the five regular solids—he believed that the particles of earth, air, fire and water had the shapes of cubes, octahedra, tetrahedra and icosahedra, while the universe itself was shaped like a dodecahedron. To these may perhaps be added Plato's belief that the distances of the sun, moon and planets were "in the proportion of the double intervals," by which he meant the sequence of integers which are powers of either 2 or 3—namely 1, 2, 3, 4, 8, 9, 27.

If any of these considerations retain any shred of validity to-day, it is the first—the universe of the theory of relativity is finite just because it is geometrical. The idea that the four elements and the universe were in any way related to the five regular solids was of course mere fancy, and the true distances of the sun, moon and planets bear absolutely no relation to Plato's numbers.

Two thousand years after Plato, Kepler spent much time and energy in trying to relate the sizes of the planetary orbits to musical intervals and geometrical constructions; perhaps he, too, hoped to discover that the orbits had been arranged by a musician or a geometer. Indeed at one time he believed he had found that the ratios of the orbits were related to the geometry of the five regular solids. If this supposed fact had been known to Plato, what a proof he might have seen in it of the geometrising propensities of the deity! Kepler himself wrote: "The intense pleasure I have received from this discovery can never be told in words." It need hardly be said that the great discovery was fallacious. Indeed our modern minds immediately dismiss it as ridiculous; we find it impossible to think of the solar system as a finished product, the same to-day as when it came from the hand of its maker; we can only think of it as something continually changing and evolving, working out its own future from its past. Yet if we can momentarily give a sufficiently mediaeval cast to our thoughts, and imagine anything so fanciful as that Kepler's conjecture should have been true, it is clear that he would have been entitled to draw some sort of inference from it. The mathematics which he had found in the universe would have been something more than he had himself put in, and he could legitimately have argued that there was inherent in the universe a mathematics additional to that which he had used to unravel its design; he might have argued, in anthropomorphic language, that his discovery suggested that the universe had been designed by a geometer. And he need no more have troubled about the criticism that the mathematics he had discovered resided merely in his own mathematical spectacles, than the angler who catches a big fish by using a little fish as bait need be worried by the comment—"Yes, but I saw you put the fish in yourself."

Let us take a more modern and less fanciful example of the same thing. Fifty years ago, when there was much discussion on the problem of communicating with Mars, it was desired to notify the supposed Martians that thinking beings existed on the planet Earth, but the difficulty was to find a language understood by both parties. The suggestion was made that the most suitable language was that of pure mathematics; it was proposed to light chains of bonfires in the Sahara, to form a diagram illustrating the famous theorem of Pythagoras, that the squares on the two smaller sides of a right-angled triangle are together equal to the square on the greatest side. To most of the inhabitants of Mars such signals would convey no meaning, but it was argued that mathematicians on Mars, if such existed, would surely recognise them as the handiwork of mathematicians on earth. In so doing, they would not be open to the reproach that they saw mathematics in everything. And it seems to me that the situation is similar, *mutatis mutandis*, with the signals from the outer world of reality which form the shadows on the walls of the cave in which we are imprisoned. We cannot interpret these as shadows cast by living actors nor as shadows cast by a machine, but the pure mathematician recognises them as representing the kind of ideas with which he is already familiar in his studies.

We could not of course draw any conclusion from this

if the concepts of pure mathematics which we find to be inherent in the structure of the universe were merely part of, or had been introduced through, the concepts of applied mathematics which we used to discover the workings of the universe. It would prove nothing if nature had merely been found to act in accordance with the concepts of applied mathematics; these concepts were specially and deliberately designed by man to fit the workings of nature. Thus it may still be objected that even our pure mathematics does not in actual fact represent a creation of our own minds so much as an effort, based on forgotten or subconscious memories, to understand the workings of nature. If so, it is not surprising that nature should be found to work according to the laws of pure mathematics. It cannot of course be denied that some of the concepts with which the pure mathematician works are taken direct from his experience of nature. An obvious instance is the concept of quantity, but this is so fundamental that it is hard to imagine any scheme of nature from which it was entirely excluded. Other concepts borrow at least something from experience; for instance multi-dimensional geometry, which clearly originated out of experience of the three dimensions of space. If, however, the more intricate concepts of pure mathematics have been transplanted from the workings of nature, they must have been buried very deep indeed in our sub-conscious minds. This very controversial possibility is one which cannot be entirely dismissed, but it is exceedingly hard to believe that such intricate concepts as a finite curved space and an expanding space can have entered into pure mathematics through any sort of unconscious or subconscious experience of the workings of the actual universe. In any event, it can hardly be disputed that nature and our conscious mathematical minds work according to the same laws. She does not model her behaviour, so to speak, on that forced on us by our whims and passions, or on that of our muscles and joints, but on that of our thinking minds. This remains true whether our minds impress their laws on nature, or she impresses her laws on us, and provides a sufficient justification for thinking of the universe as being of mathematical design. Lapsing back again into the crudely anthropomorphic language we have already used, we may say that we have already considered with disfavour the possibility of the universe having been planned by a biologist or an engineer; from the intrinsic evidence of his creation, the Great Architect of the Universe now begins to appear as a pure mathematician.

Personally I feel that this train of thought may, very tentatively, be carried a stage farther, although it is difficult to express it in exact words, again because our mundane vocabulary is circumscribed by our mundane experience. The terrestrial pure mathematician does not concern himself with material substance, but with pure thought. His creations are not only created by thought but consist of thought, just as the creations of the engineer consist of engines. And the concepts which now prove to be fundamental to our understanding of nature—a space which is finite; a space which is empty, so that one point differs from another solely in the properties of the space itself; four-dimensional, seven- and more dimensional spaces; a space which for ever expands; a sequence of events which follows the laws of probability instead of the law of causation—or, alternately, a sequence of events which can only be fully and consistently described by going outside space and time—all these concepts seem to my mind to be structures of pure thought, incapable of realisation in any sense which would properly be described as material.

For instance, anyone who has written or lectured on the finiteness of space is accustomed to the objection that the concept of a finite space is self-contradictory and nonsensical. If space is finite, our critics say, it must be possible to go out beyond this finite space, and what can we possibly find beyond it except more space, and so on *ad infinitum*?—which proves that space cannot be finite. And again, they say, if space is expanding, what can it possibly expand into, if not into more space?—which again proves that what is expanding can only be a part of space, so that the whole of space cannot expand.

The twentieth-century critics who make these comments are still in the state of mind of the nineteenth-century scientists; they take it for granted that the universe must admit of material representation. If we grant their premisses, we must, I think, also grant their conclusion—that we are talking nonsense—for their logic is irrefutable. But modern science cannot possibly grant their conclusion; it insists on the finiteness of space at all costs. This of course means that we must deny the premisses which our critics unknowingly assume. The universe cannot admit of material representation, and the reason, I think, is that it has become a mere mental concept.

It is the same, I think, with other more technical concepts, typified by the "exclusion principle," which seem to imply a sort of "action-at-a-distance" in both space and time—as though every bit of the universe knew what other distant bits were doing, and acted accordingly. To my mind, the laws which nature obeys are less suggestive of those which a machine obeys in its motion than of those which a musician obeys in writing a fugue, or a poet in composing a sonnet. The motions of electrons and atoms do not resemble those of the parts of a locomotive so much as those of the dancers in a cotillion. And if the "true essence of substances" is for ever unknowable, it does not matter whether the cotillion is danced at a ball in real life, or on a cinematograph screen, or in a story of Boccaccio. If all this is so, then the universe can be best pictured, although still very imperfectly and inadequately, as consisting of pure thought, the thought of what, for want of a wider word, we must describe as a mathematical thinker.

And so we are led into the heart of the problem of the relation between mind and matter. Atomic disturbances in the distant sun cause it to emit light and heat. After "travelling through the ether" for eight minutes, some of this radiation may fall on our eyes, causing a disturbance on the retina, which travels along the optic nerve to the brain. Here it is perceived as a sensation by the mind; this sets our thoughts in action and results in, let us say, poetic thoughts about the sunset. There is a continuous chain, A, B, C, D...X, Y, Z, connecting A the poetic thought—through B the thinking mind, C the brain, D the optic nerve, and so on—with Z the atomic disturbance in the sun. The thought A results from the distant disturbance Z, just as the ringing of a bell results from pulling a distant bell-rope. We can understand how pulling a material rope can cause a material bell to ring, because there is a material connection all the way. But it is far less easy to see how a disturbance of material atoms can cause a poetic thought to originate, because the two are so entirely dissimilar in nature.

For this reason, Descartes insisted that there could be no

possible connection between mind and matter. He believed they were two entirely distinct kinds of entity, the essence of matter being extension in space, and that of mind being thought. And this led him to maintain that there were two distinct worlds, one of mind and one of matter, running, so to speak, independent courses on parallel rails without ever meeting.

Berkeley and the idealist philosophers agreed with Descartes that if mind and matter were fundamentally of different natures they could never interact. But they insisted that they continually do interact. Therefore, they argued, matter must be of the same nature as mind, so that, in the terminology of Descartes, the essence of matter must be thought rather than extension. Expressed in detail, their contention was that causes must be essentially of the same nature as their effects; if *B* on our chain produces *A*, then *B* must be of the same essential nature as *A*, and *C* as *B*, and so on. Thus *Z* also must be of the same essential nature as *A*. Now the only links of the chain of which we have any *direct* knowledge are our own thoughts and sensations *A*, *B*; we know of the existence and nature of the remote links *X*, *Y*, *Z* only by inference—from the effects they transmit to our minds through our senses. Berkeley, maintaining that the unknown distant links *X*, *Y*, *Z*, must be of the same nature as the known near links *A*, *B*, argued that they must be of the nature of thoughts or ideas, "since after all there is nothing like an idea except an idea." A thought or idea cannot, however, exist without a mind in which to exist. We may say an object exists in our minds while we are conscious of it, but this will not account for its existence during the time we are not conscious of it. The planet Pluto, for instance, was in existence long before any human mind suspected it, and was recording its existence on photographic plates long before any human eye saw it. Considerations such as these led Berkeley to postulate an Eternal Being, in whose mind all objects existed. And so, in the stately and sonorous diction of a bygone age, he summed up his philosophy in the words:

> All the choir of heaven and furniture of earth, in a word all those bodies which compose the mighty frame of the world, have not any substance without the mind....So long as they are not actually perceived by me, or do not exist in my mind, or that of any other created spirit, they must either have no existence at all, or else subsist in the mind of some Eternal Spirit.

Modern science seems to me to lead, by a very different road, to a not altogether dissimilar conclusion. Biology, studying the connection between the earlier links of the chain, *A*, *B*, *C*, *D*, seems to be moving towards the conclusion that these are all of the same general nature. This is occasionally stated in the specific form that, as biologists believe *C*, *D* to be mechanical and material, *A*, *B* must also be mechanical and material, but apparently there would be at least equal warrant for stating it in the form that as *A*, *B* are mental, *C*, *D* must also be mental. Physical science, troubling little about *C*, *D*, proceeds directly to the far end of the chain; its business is to study the workings of *X*, *Y*, *Z*. And, as it seems to me, its conclusions suggest that the end links of the chain, whether we go to the cosmos as a whole or to the innermost structure of the atom, are of the same nature as *A*, *B*—of the nature of pure thought; we are led to the conclusions of Berkeley, but we reach them from the other end. Because of this, we come upon the last of Berkeley's three alternatives first, and the others appear unimportant by comparison. It does not matter whether objects "exist in my mind, or that of any other created spirit" or not; their objectivity arises from their subsisting "in the mind of some Eternal Spirit."

This may suggest that we are proposing to discard realism entirely, and enthrone a thoroughgoing idealism in its place. Yet this, I think, would be too crude a statement of the situation. If it is true that the "real essence of substances" is beyond our knowledge, then the line of demarcation between realism and idealism becomes very blurred indeed; it becomes little more than a relic of a past age in which reality was believed to be identical with mechanism. Objective realities exist, because certain things affect your consciousness and mine in the same way, but we are assuming something we have no right to assume if we label them as either "real" or "ideal." The true label is, I think, "mathematical," if we can agree that this is to connote the whole of pure thought, and not merely the studies of the professional mathematician. Such a label does not imply anything as to what things are in their ultimate essence, but merely something as to how they behave.

The label we have selected does not of course relegate matter into the category of hallucination or dreams. The material universe remains as substantial as ever it was, and this statement must, I think, remain true through all changes of scientific or philosophical thought.

For substantiality is a purely mental concept measuring the direct effect of objects on our sense of touch. We say that a stone or a motor-car is substantial, while an echo or a rainbow is not. This is the ordinary definition of the word, and it is a mere absurdity, a contradiction in terms, to say that stones and motor-cars can in any way become insubstantial, or even less substantial, because we now associate them with mathematical formulae and thoughts, or kinks in empty space, rather than with crowds of hard particles. Dr Johnson is reported to have expressed his opinion on Berkeley's philosophy by dashing his foot against a stone and saying: "No, Sir, I disprove it thus." This little experiment had of course not the slightest bearing on the philosophical problem it claimed to solve; it merely verified the substantiality of matter. And, however science may progress, stones must always remain substantial bodies, just because they and their class form the standard by which we define the quality of substantiality.

It has been suggested that the lexicographer might really have disproved the Berkeleian philosophy if he had chanced to kick, not a stone but a hat, in which some small boy had surreptitiously placed a brick; we are told that "the element of surprise is sufficient warrant for external reality," and that "a second warrant is permanence with change—permanence in your own memory, change in externality." This of course merely disproves the solipsist error of "all this is a creation of my own mind, and exists in no other mind," but it is hard to do anything in life which does not disprove this. The argument from surprise, and from new knowledge in general, is powerless against the concept of a universal mind of which your mind and mine, the mind which surprises and that which is surprised, are units or even excrescences. Each individual brain cell cannot be acquainted with all the thoughts which are passing through the brain as a whole.

Yet the fact that we possess no absolute extraneous standard against which to measure substantiality does not preclude our saying that two things have the same degree, or different degrees, of substantiality. If I dash my foot against a stone in my dreams, I shall probably waken up with a pain in my foot, to discover that the stone of my dreams was literally a creation of my mind and of mine alone, prompted by a nerve-impulse originating in my foot. This stone may typify the category of hallucinations or dreams; it is clearly less substantial than that which Johnson kicked. Creations of an individual mind may reasonably be called less substantial than creations of a universal mind. A similar distinction must be made between the space we see in a dream and the space of everyday life; the latter, which is the same for us all, is the space of the universal mind. It is the same with time, the time of waking life, which flows at the same even rate for us all, being the time of the universal mind. Again we may think of the laws to which phenomena conform in our waking hours, the laws of nature, as the laws of thought of a universal mind. The uniformity of nature proclaims the self-consistency of this mind.

This concept of the universe as a world of pure thought throws a new light on many of the situations we have encountered in our survey of modern physics. We can now see how the ether, in which all the events of the universe take place, could reduce to a mathematical abstraction, and become as abstract and as mathematical as parallels of latitude and meridians of longitude. We can also see why energy, the fundamental entity of the universe, had again to be treated as a mathematical abstraction—the constant of integration of a differential equation.

The same concept implies of course that the final truth about a phenomenon resides in the mathematical description of it; so long as there is no imperfection in this our knowledge of the phenomenon is complete. We go beyond the mathematical formula at our own risk; we may find a model or picture which helps us to understand it, but we have no right to expect this, and our failure to find such a model or picture need not indicate that either our reasoning or our knowledge is at fault. The making of models or pictures to explain mathematical formulae and the phenomena they describe, is not a step towards, but a step away from, reality; it is like making graven images of a spirit. And it is as unreasonable to expect these various models to be consistent with one another as it would be to expect all the statues of Hermes, representing the god in all his varied activities—as messenger, herald, musician, thief, and so on—to look alike. Some say that Hermes is the wind; if so, all his attributes are wrapped up in his mathematical description, which is neither more nor less than the equation of motion of a compressible fluid. The mathematician will know how to pick out the different aspects of this equation which represent the conveying and announcing of messages, the creation of musical tones, the blowing away of our papers, and so forth. He will hardly need statues of Hermes to remind him of them, although, if he is to rely on statues, nothing less than a whole row, all different, will suffice. All the same, some mathematical physicists are still busily at work making graven images of the concepts of the wave-mechanics.

In brief, a mathematical formula can never tell us what a thing is, but only how it behaves; it can only specify an object through its properties. And these are unlikely to coincide *in toto* with the properties of any single macroscopic object of our everyday life.

This point of view brings us relief from many of the difficulties and apparent inconsistencies of present-day physics. We need no longer discuss whether light consists of particles or waves; we know all there is to be known about it if we have found a mathematical formula which accurately describes its behaviour, and we can think of it as either particles or waves according to our mood and the convenience of the moment. On our days of thinking of it as waves, we may if we please imagine an ether to transmit the waves, but this ether will vary from day to day; we have seen how it will vary each time our speed of motion varies. In the same way, we need not discuss whether the wave-system of a group of electrons exists in a three-dimensional space, or in a many-dimensional space, or not at all. It exists in a mathematical formula; this, and nothing else, expresses the ultimate reality, and we can picture it as representing waves in three, six or more dimensions whenever we so please. We can also interpret it as not representing waves at all; in so doing we shall be following Heisenberg and Dirac. It is generally simplest to interpret it as representing waves in a space having three dimensions for each electron, just as it is simplest to interpret the macroscopic universe as an array of objects in three dimensions only, and its phenomena as an array of events in four dimensions, but none of these interpretations possesses any unique or absolute validity.

On this view, we need find no mystery in the nature of the rolling contact of our consciousness with the empty soap-bubble we call space-time (p. 105), for it reduces merely to a contact between mind and a creation of mind—like the reading of a book, or listening to music. It is probably unnecessary to add that, on this view of things, the apparent vastness and emptiness of the universe, and our own insignificant size therein, need cause us neither bewilderment nor concern. We are not terrified by the sizes of the structures which our own thoughts create, nor by those that others imagine and describe to us. In du Maurier's story, Peter Ibbetson and the Duchess of Towers continued to build vast dream-palaces and dream-gardens of ever-increasing size, but felt no terror at the size of their mental creations. The immensity of the universe becomes a matter of satisfaction rather than awe; we are citizens of no mean city. Again, we need not puzzle over the finiteness of space; we feel no curiosity as to what lies beyond the four walls which bound our vision in a dream.

It is the same with time, which, like space, we must think of as of finite extent. As we trace the stream of time backwards, we encounter many indications that, after a long enough journey, we must come to its source, a time before which the present universe did not exist. Nature frowns upon perpetual motion machines and it is *à priori* very unlikely that her universe will provide an example, on the grand scale, of the mechanism she abhors. And a detailed consideration of nature confirms this. The science of thermodynamics explains how everything in nature passes to its final state by a process which is designated the "increase of entropy." Entropy must for ever increase: it cannot stand still until it has increased so far that it can increase no further. When this stage is reached, further progress will be impossible, and the universe will be dead. Thus, unless this whole branch of science is wrong, nature permits herself, quite literally, only two alternatives, progress and death: the only standing still she permits is in the stillness of the grave.

Some scientists, although not, I think, very many, would dissent from this last view. While they do not dispute that the present stars are melting away into radiation, they maintain that, somewhere out in the remote depths of space, this radiation may be reconsolidating itself again into matter. A new heaven and a new earth may, they suggest, be in process of being built, not out of the ashes of the old, but out of the radiation set free by the combustion of the old. In this way they advocate what may be described as a cyclic universe; while it dies in one place the products of its death are busy producing new life in others.

This concept of a cyclic universe is entirely at variance with the well-established principle of the second law of thermodynamics, which teaches that entropy must for ever increase, and that cyclic universes are impossible in the same way, and for much the same reason, as perpetual motion machines are impossible. That this law may fail under astronomical conditions of which we have no knowledge is certainly conceivable, although I imagine the majority of serious scientists consider it very improbable. There is of course no denying that the concept of a cyclic universe is far the more popular of the two. Most men find the final dissolution of the universe as distasteful a thought as the dissolution of their own personality, and man's strivings after personal immortality have their macroscopic counterpart in these more sophisticated strivings after an imperishable universe.

The more orthodox scientific view is that the entropy of the universe must for ever increase to its final maximum value. It has not yet reached this: we should not be thinking about it if it had. It is still increasing rapidly, and so must have had a beginning; there must have been what we may describe as a "creation" at a time not infinitely remote.

If the universe is a universe of thought, then its creation

must have been an act of thought. Indeed the finiteness of time and space almost compel us, of themselves, to picture the creation as an act of thought; the determination of the constants such as the radius of the universe and the number of electrons it contained imply thought, whose richness is measured by the immensity of these quantities. Time and space, which form the setting for the thought, must have come into being as part of this act. Primitive cosmologies pictured a creator working in space and time, forging sun, moon and stars out of already existent raw material. Modern scientific theory compels us to think of the creator as working outside time and space, which are part of his creation, just as the artist is outside his canvas. It accords with the conjecture of Augustine: "Non in tempore, sed cum tempore, finxit Deus mundum." Indeed, the doctrine dates back as far as Plato:

> Time and the heavens came into being at the same instant, in order that, if they were ever to dissolve, they might be dissolved together. Such was the mind and thought of God in the creation of time.

And yet, so little do we understand time that perhaps we ought to compare the whole of time to the act of creation, the materialisation of the thought.

It may be objected that our whole argument is based on the assumption that the present mathematical interpretation of the physical world is in some way unique, and will prove to be final. To resume our metaphor, it may be said that to describe the reality as a game of chess is only a convenient fiction: other fictions might describe the motions of the shadows equally well. The answer is that, so far as our present knowledge goes, other fictions would not describe them so fully, so simply, or so adequately. The man who does not play chess says: "A piece of white wood, carved to look rather like a horse's head stuck on a pedestal, was taken from the bottom square next but one to the right-hand corner and moved to..." and so on. The chess-player says, "White: Kt to KB3," and his account not only explains the move fully and briefly, but also relates it to a larger scheme of things. In science, so long as our knowledge remains incomplete, the simplest explanation carries conviction in proportion to its simplicity. And it has merit beyond that of mere simplicity: it has the highest probability of being the true explanation. Thus while it must be fully admitted that the mathematical explanation may prove neither to be final nor the simplest possible, we can unhesitatingly say that it is the simplest and most complete so far found, so that, relative to our present knowledge, it has the greatest chance of being the explanation which lies nearest to the truth.

Some readers may not assent to this, on the grounds that the present-day mathematical interpretation of nature is likely to prove a mere half-way house to a new mechanical interpretation. Our modern minds have, I think, a bias towards mechanical interpretations. Part may be due to our early scientific training; part perhaps to our continually seeing everyday objects behaving in a mechanical way, so that a mechanical explanation looks natural and is easily comprehended. Yet in a completely objective survey of the situation, the outstanding fact would seem to be that mechanics has already shot its bolt and has failed dismally, on both the scientific and philosophical side. If anything is destined to replace mathematics, there would seem to be specially long odds against it being mechanics.

It is too often overlooked that we can only discuss these

questions in terms of probabilities. The man of science is accustomed to the reproach that he changes his views all the time, with the accompanying implication that what he says need not be taken too seriously. It is no true reproach that in exploring the river of knowledge he occasionally goes down a backwater instead of continuing along the main stream; no explorer can be sure that a backwater is such, and nothing more, until he has been down it. What is more serious, and beyond the control of the explorer, is that the river is a winding one, flowing now east, now west. At one moment the explorer says: "I am going downstream, and, as I am going towards the west, the ocean which is reality seems most likely to lie in the westerly direction." And later, when the river has turned east, he says: "It now looks as though reality is in the east." No scientist who has lived through the last thirty years is likely to be too dogmatic either as to the future course of the stream or as to the direction in which reality lies: he knows from his own experience how the river not only for ever broadens but also repeatedly winds, and, after many disappointments, he has given up thinking at every turn that he is at last in the presence of the

murmurs and scents of the infinite sea.

With this caution in mind, it seems at least safe to say that the river of knowledge has made a sharp bend in the last few years. Thirty years ago, we thought, or assumed, that we were heading towards an ultimate reality of a mechanical kind. It seemed to consist of a fortuitous jumble of atoms, which was destined to perform meaningless dances for a time under the action of blind purposeless forces, and then fall back to form a dead world. Into this wholly mechanical world, through the play of the same blind forces, life had stumbled by accident. One tiny corner at least, and possibly several tiny corners, of this universe of atoms had chanced to become conscious for a time, but was destined in the end, still under the action of blind mechanical forces, to be frozen out and again leave a lifeless world.

To-day there is a wide measure of agreement, which on the physical side of science approaches almost to unanimity, that the stream of knowledge is heading towards a non-mechanical reality; the universe begins to look more like a great thought than like a great machine. Mind no longer appears as an accidental intruder into the realm of matter; we are beginning to suspect that we ought rather to hail it as the creator and governor of the realm of matter—not of course our individual minds, but the mind in which the atoms out of which our individual minds have grown exist as thoughts.

The new knowledge compels us to revise our hasty first impressions that we had stumbled into a universe which either did not concern itself with life or was actively hostile to life. The old dualism of mind and matter, which was mainly responsible for the supposed hostility, seems likely to disappear, not through matter becoming in any way more shadowy or insubstantial than heretofore, or through mind becoming resolved into a function of the working of matter, but through substantial matter resolving itself into a creation and manifestation of mind. We discover that the universe shews evidence of a designing or controlling power that has something in common with our own individual minds—not, so far as we have discovered, emotion, morality, or aesthetic appreciation, but the tendency to think in the way which, for want of a better word, we describe as mathematical. And while much in it may be hostile to the material appendages of life, much also is akin to the fundamental activities of life; we are not so much strangers or intruders in the universe as we at first thought. Those inert atoms in the primaeval slime which first began to foreshadow the attributes of life were putting themselves more, and not less, in accord with the fundamental nature of the universe.

So at least we are tempted to conjecture to-day, and yet who knows how many more times the stream of knowledge may turn on itself? And with this reflection before us, we may well conclude by adding, what might well have been interlined into every paragraph, that everything that has been said, and every conclusion that has been tentatively put forward, is quite frankly speculative and uncertain. We have tried to discuss whether present-day science has anything to say on certain difficult questions, which are perhaps set for ever beyond the reach of human understanding. We cannot claim to have discerned more than a very faint glimmer of light at the best; perhaps it was wholly illusory, for certainly we had to strain our eyes very hard to see anything at all. So that our main contention can hardly be that the science of to-day has a pronouncement to make, perhaps it ought rather to be that science should leave off making pronouncements: the river of knowledge has too often turned back on itself.

INDEX